Advances in Science, Technology & Innovation

IEREK Interdisciplinary Series for Sustainable Development

Advances in Science, Technology & Innovation (ASTI) is a series of peer-reviewed books based on important emerging research that redefines the current disciplinary boundaries in science, technology and innovation (STI) in order to develop integrated concepts for sustainable development. It not only discusses the progress made towards securing more resources, allocating smarter solutions, and rebalancing the relationship between nature and people, but also provides in-depth insights from comprehensive research that addresses the **17 sustainable development goals (SDGs)** as set out by the UN for 2030.

The series draws on the best research papers from various IEREK and other international conferences to promote the creation and development of viable solutions for a **sustainable future and a positive societal** transformation with the help of integrated and innovative science-based approaches. Including interdisciplinary contributions, it presents innovative approaches and highlights how they can best support both economic and sustainable development, through better use of data, more effective institutions, and global, local and individual action, for the welfare of all societies.

The series particularly features conceptual and empirical contributions from various interrelated fields of science, technology and innovation, with an emphasis on digital transformation, that focus on providing practical solutions to **ensure food, water and energy security to achieve the SDGs**. It also presents new case studies offering concrete examples of how to resolve sustainable urbanization and environmental issues in different regions of the world.

The series is intended for professionals in research and teaching, consultancies and industry, and government and international organizations. Published in collaboration with IEREK, the Springer ASTI series will acquaint readers with essential new studies in STI for sustainable development.

ASTI series has now been accepted for Scopus (September 2020). All content published in this series will start appearing on the Scopus site in early 2021.

Abdul Hai Alami
Editor

Sustainable Additive Technologies and Innovative Manufacturing Techniques

Springer

Editor
Abdul Hai Alami
Department of Sustainable
and Renewable Energy Engineering
University of Sharjah
Sharjah, United Arab Emirates

ISSN 2522-8714 ISSN 2522-8722 (electronic)
Advances in Science, Technology & Innovation
IEREK Interdisciplinary Series for Sustainable Development
ISBN 978-3-031-73656-8 ISBN 978-3-031-73657-5 (eBook)
https://doi.org/10.1007/978-3-031-73657-5

This Springer imprint is published by the registered company Springer Nature Switzerland AG
The registered company address is: Gewerbestrasse 11, 6330 Cham, Switzerland

To the three dimensions of my life; Seif, Suleiman and Sultan. For adding to and sintering my existence. This also goes to my wonderful coauthors M. Ayoub, A. Yasin, Ayman, Adnan, Siren, Shamma, Haya, Fabrizia and Fatima. I remain appreciative and grateful.

Foreword

It is with great pleasure and a deep sense of admiration that I introduce this book on additive manufacturing, authored by my long-time friend, teacher and colleague, Prof. Abdul Hai Alami. Over the past decade, I have made significant award-winning contributions to the field of materials science, particularly in advancing the understanding and application of cellular materials and lattices through innovative additive manufacturing techniques. As an expert in Additive and Advanced Manufacturing, and as a frequent reviewer for top-ranked journals in Manufacturing such as Additive Manufacturing and Additive Manufacturing Letters, I believe that it is time to have a comprehensive guide that seamlessly integrates sustainable practices with cutting-edge additive manufacturing techniques, fostering both technical expertise and innovative thinking. This book is precisely that, a guide for those who want to delve into Additive Manufacturing and avoid all the noise and hype associated with it. I have known Prof. Alami for over 15 years and have had the privilege of witnessing his dedication to education and his passion for fostering innovation in his students. I had the privilege of testing it firsthand as a student and later as a collaborator. His unique approach to teaching is beautifully captured in this book, which serves as a testament to his belief in the transformative power of learning. I know firsthand how Alami's teaching philosophy is rooted in a deep understanding of both what students do and how they think. He has always emphasized the importance of equipping students not just with technical knowledge, but with the ability to think critically and creatively. This book is a manifestation of that philosophy, offering a balanced mix of declarative knowledge and practical, functioning know-how. This can be felt straight from the first chapter which sets the mind to accept that design is not meant only for those who are experienced and talented but it can actually be a thinking process that, with the right tools, can transfer a concept all the way to a product. Alami's commitment to sustainability is evident throughout the chapters of this book. As a faculty member in sustainable and renewable engineering at Sharjah University, his work consistently reflects his dedication to integrating environmentally conscious practices into the curriculum. His focus on sustainability is not just a trend but a core principle that informs his teaching and research. This is clear in how he presents the material, always with an eye towards the future and its impact on our planet. Reading through the chapters, you will find a careful blend of theory and practice. The last chapter manifests the "do it yourself" approach in the most elegant way. This chapter equips the reader with the needed know-how to indulge in the experience of forming metals into complex components in a backyard. Through the merging of fairly new Additive Manufacturing techniques and the ancient casting techniques to fabricate metallic components that would ideally require a factory or machine shop setup or advanced expensive metal additive manufacturing techniques. This book will be a cornerstone for engineering students, hobbyists, and business owners. I am looking forward to having it as part of my collection.

Dr. Oraib Al-Ketan
Research Instrumentation Scientist
Advanced Manufacturing Core Facility
New York University Abu Dhabi
Abu Dhabi, United Arab Emirates
e-mail: oraib.alketan@nyu.edu

Preface

The global population is currently facing a rare phenomenon that never existed before. Technological advancement is accelerating at such a pace that a new technology is emerging every fortnight if not sooner. With the digital revolution that was amplified in 2007 and with the introduction of the iPhone and other smart devices, the world has been freely and independently interconnected. A product that is available in Europe would be coveted by consumers in China in a matter of days. This has created a major strain on supply chains to provide products to hungry consumers, and manufacturers have been racing to meet the demand and make more profit than their competitors.

With the consumer culture being the norm, balancing resource harvesting with sustainability as well as environmental consciousness has become a serious challenge. More production means more resources needed, more emissions, more pollution and more waste. This has necessitated a revision of the manufacturing paradigm to balance the need for increased production while being considerate of any adverse impact on ecosystems and climate change.

This book takes the reader on a journey to do just that. It starts with translating consumer requirements into design specifications and constraints while maintaining a certain quality and an edge over the competition. This can be done by systemizing the transition from concept to CAD to product, where the manufacturing process is assigned and identified. The challenge with conventional manufacturing is how energy-intensive and wasteful it is. Thus, the main intention of this book is to introduce sustainable additive manufacturing technologies that would result in better resource utilization, less waste and energy efficiency.

The book is useful for the novice and the seasoned manufacturing enthusiast, as it documents the many paths available to arrive at the desired product. From a brief discussion on conventional manufacturing to a real case study of lost wax casting, the book is sure to appeal to the reader and be a useful guide on sustainable additive manufacturing and the role of innovation in an industry that is growing in size and capabilities every day.

Abdul Hai Alami, Ph.D.
Professor and Chair
Department of Sustainable
and Renewable Energy Engineering
University of Sharjah
Sharjah, United Arab Emirates
e-mail: aalalami@sharjah.ac.ae

Contents

About the Editor

Abdul Hai Alami currently a Professor at and Chairman of the Sustainable and Renewable Energy Engineering Program at the University of Sharjah (2012–Present). Professor Alami has received his Ph.D. from Queen's University in Kingston, ON, Canada, in 2006. Since then, he has held positions at the Mechanical Engineering Department at the Hashemite University in Jordan (2006–2010) and Al Ain in UAE (2010–2012). Professor Alami has been among the world's top 2% scientists according to Stanford University Scientists Ranking since 2020.

The areas of interest of Prof. Alami are the synthesis and analysis of nano-materials for various energy conversion and storage applications, as well as novel ways of mechanical energy storage (CAES, CGES, and buoyancy force). Professor Alami has authored more than 200 publications in highly reputed journals, as well as books on energy storage technologies and solar photovoltaic manufacturing.

List of Figures

Additive Manufacturing Technologies

Additive Manufacturing of Sustainable Functional Parts

Recent Advances in Additive Manufacturing Technologies

Do-It-Yourself Project: Sand Casting and Lost PLA/Wax

Road to Best Design

Abdul Hai Alami, Mohamad Ayoub, Ahmad Yasin, Fabrizia Devito, Shamma Alabdulla, Ayman Mdallal, and Adnan Alashkar

Abstract

Going from concept to product requires many stops in between. Translating end-user requirements into product specifications should not depend solely on the experience and talent of individuals at the design office. This chapter introduces a repeatable methodology with which any design idea can be rationalized with certain constraints that may come from materials, costs, process selection, or utilization of sustainable manufacturing processes or recyclable materials. Although a talented designer provides a unique signature to their products, complete reliance on it jeopardizes the sequential nature of sustainable production. To overcome this talent-related constraint, a methodological approach such as House of Quality (HoQ) is used to arrive at the best design.

1 Introduction

Talent management is a critical factor in the success of any design and manufacturing process and is particularly important in additive manufacturing. The ability to attract and retain talented individuals enables organizations to harness advanced technical knowledge, foster innovation, and create superior products that stand out in the market. As the field of additive manufacturing continues to evolve, the strategic management of talent will remain a critical factor in achieving sustainable growth and competitive advantage.

Product manufacturing is in no way a straightforward task. Various factors are involved, including the client's needs and the manufacturer's ability to cater to them and tailor the product in a way that minimizes costs and retains quality. However, this does not mean that good products are a monopoly of talented manufacturers or designers who can overcome trade-offs in designs based on experience and intuition. There are proven methodological approaches that yield satisfactory product design results based on available requirements and user inputs.

Quality Function Deployment (QFD) is a critical tool to ensure the systematic translation of client requirements into precise technical specifications across all stages of the design and manufacturing processes. These stages harbor a wide spectrum of activities, including marketing strategies, strategic planning, product design and development, prototype evaluation, manufacturing process development, and mass production. In this chapter, talent's role in the design process, as well as House of Quality, which is a part of QFD methodology, and life cycle analysis that considers the economic and environmental impacts of products, are

A. H. Alami (✉) · M. Ayoub · A. Yasin
Sustainable and Renewable Energy Engineering Department, University of Sharjah, Sharjah, 27272, United Arab Emirates
e-mail: aalalami@sharjah.ac.ae

M. Ayoub
e-mail: mohamad.ayoub@sharjah.ac.ae

A. Yasin
e-mail: ahmad.yasin@sharjah.ac.ae

F. Devito
Department of Mechanical, Mathematics and Management (DMMM), Polytechnical University of Bari, Bari, Italy
e-mail: fabrizia.devito@poliba.it; fabrizia.devito@iusspavia.it

Department of Sciences, Technologies and Society, University School for Advanced Studies Pavia - IUSS Pavia, Pavia, Italy

S. Alabdulla
Mechanical Engineering Department, American University of Sharjah, 26666 Sharjah, UAE
e-mail: g00070854@aus.edu

A. Mdallal
Sustainable Energy & Power Systems Research Centre, RISE, University of Sharjah, Sharjah, 27272, United Arab Emirates
e-mail: Ayman.Mdallal@sharjah.ac.ae

A. Alashkar
Ph.D. Program in Materials Science and Engineering, American University of Sharjah, Sharjah, 26666, United Arab Emirates
e-mail: b00028197@aus.edu

A. H. Alami (ed.), *Sustainable Additive Technologies and Innovative Manufacturing Techniques*, Advances in Science, Technology & Innovation, https://doi.org/10.1007/978-3-031-73657-5_1

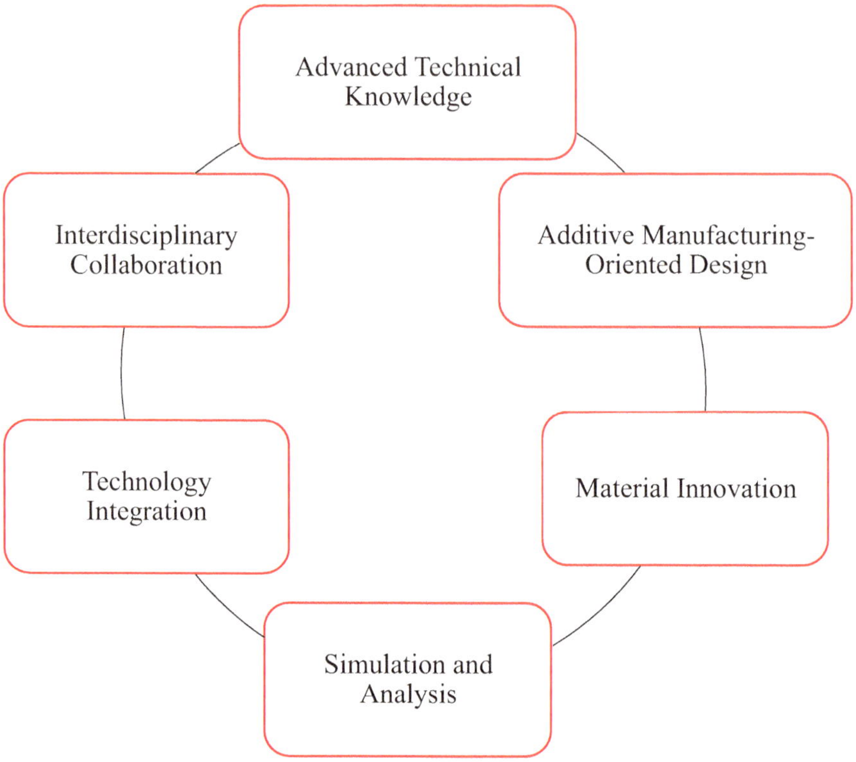

Fig. 1 Areas of additive manufacturing key area of talent

presented within the context of additive manufacturing. The House of Quality approach will be explained later in this chapter.

2 Talent Role in the Design Process

Talent is recognized as effectively utilizing one's cognitive abilities, emotions, and behavioral skills in a productive and innovative manner. Talented individuals are characterized by their inherent competitiveness, determination, and accountability. In the context of human resources, talent is often described as the individual ability to exceed their potential, consistently delivering superior performance and innovative solutions (Alparslan, 2020). Moreover, talent plays a fundamental role in the design process by infusing creativity, innovation, and expertise into every stage of development. By deeply understanding consumer needs and preferences, talented personnel contribute significantly to creating products that differentiate them in a competitive market.

Given the critical importance of talent to the success of product conception, design and manufacture, businesses must develop robust talent management systems to effectively attract, retain, and nurture skilled employees. This involves a comprehensive approach consisting of staffing strategies, talent evaluation, unconventional recruiting processes, active training and development activities, retention initiatives, and eventually succession planning (Alparslan, 2020). Each of these elements is vital for ensuring that the right talent is placed in the right position at the right time, thereby maximizing organizational effectiveness and innovation.

In the specialized field of additive manufacturing technologies, acquiring and managing talent is of significant importance due to the complexity and rapid evolution of this domain. Talent in additive manufacturing significantly influences the design process through several key areas, as shown in Fig. 1.

3 Process Design

Nowadays, clients approach design facilities seeking specific solutions that have to be customized according to their needs and budget. Businesses usually offer a variety of options of comparable products and services for clients to choose from. This diversity of solutions and options may cause confusion among clients and requires the implementation of robust methodologies to ensure that customer preferences and requirements are accurately addressed. Quality Function Deployment (QFD) is a critical tool in this context, which facilitates the systematic translation of client requirements into precise technical specifications across all stages of the design and manufacturing processes. These stages consist of a broad spectrum of activities, including marketing strategies, strategic planning, product design and development, prototype evaluation, manufacturing process selection,

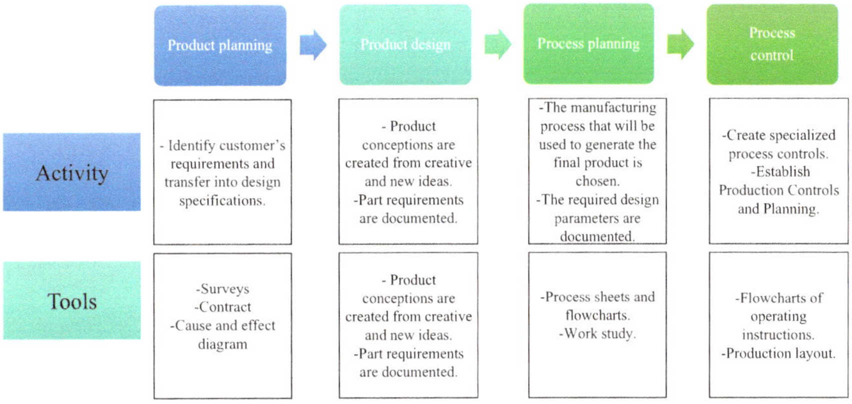

Fig. 2 The four phases in QFD

minimum order of quantity, and market introduction. QFD operates as a decision-making framework that integrates various elements such as technological advancements, internal organizational policies, and explicit client demands. The objective is to develop a product that not only meets but preferably exceeds customer expectations in terms of control, appeal, usability, and cost-effectiveness. This integrative approach ensures that every aspect of the product lifecycle is aligned with the overarching goal of customer satisfaction and retention. The decision-making technology underpinning this process is fundamentally represented by QFD (Zare Mehrjerdi, 2010) (Fig. 2).

Generally, the four phases involved in this method with their recommended tools are illustrated in Fig. 3. Product planning resembles the first step in which the voice of customers (VOC) is taken into consideration before the product design phase. The following step corresponds to identifying the significant components and requirements. The following phase, which is process planning, is based on the obtained specifications and involves designing the manufacturing process. The final phase before production is process control, testing, and inspection (Fig. 2).

4 House of Quality Approach: QFD Implementation

The strategic approach used to implement QFD for additive manufacturing is illustrated in Fig. 3. The first step is data collection where the type of customers is identified and the needs of the customer "Voice of the customer" in the materials, costs, manufacturing, and quality are considered. The collected data are then organized and used to establish the link between customer requirements and design constraints. To simplify the analysis of the factors that are present, a relationship matrix is developed, which depicts how design decisions influence the fulfillment of each customer's need. We attempt to gain a response to the question: "To what extent might the technical aspects of our product impact the quality expectations of customers for every factor in the matrix?" (Dadić et al., 2016). The House of Quality (HoQ) system is a visual methodology to develop and is a layout of all aspects of technological iteration and quality that are intertwined in the production and design of a new product. It is constructed by the QFD team from customer feedback to evaluate the best design and engineering choices for the new product.

The highly graphical aspect of the matrix allows the work team to instantaneously have a view of the strengths and weaknesses that the technical departments will have to address. Once the product's main characteristics and areas for improvement have been established, these are formalized graphically in the matrix. Each prototyping or sample product should be accompanied by an analysis based on the HoQ in order to provide immediacy on the design difficulties and engineering compromises to be made.

The concept of HoQ is a well-established practice adopted by many industrial and commercial companies to constitutionalize the concept-to-product flow of information and design constraints. Figure 4 shows the general structure of the HoQ (House of Quality | Explanation with Example—IONOS, n.d.), which can be summed up as follows:

- Customer Requirements: To ascertain precisely what a customer would want in a product, it is possible to conduct customer surveys or interviews.
- Competitive Evaluation: This part consists of a relationship matrix that assesses the relative performance of our product compared to that of the competitors.
- Product Characteristics: This section is set on top of the relationship matrix and pertains to specific product characteristics. These are technical parameters, material specifications, and potential manufacturing procedures.
- Relationship Matrix: This signifies the level of correlation between product characteristics and customer requirements.

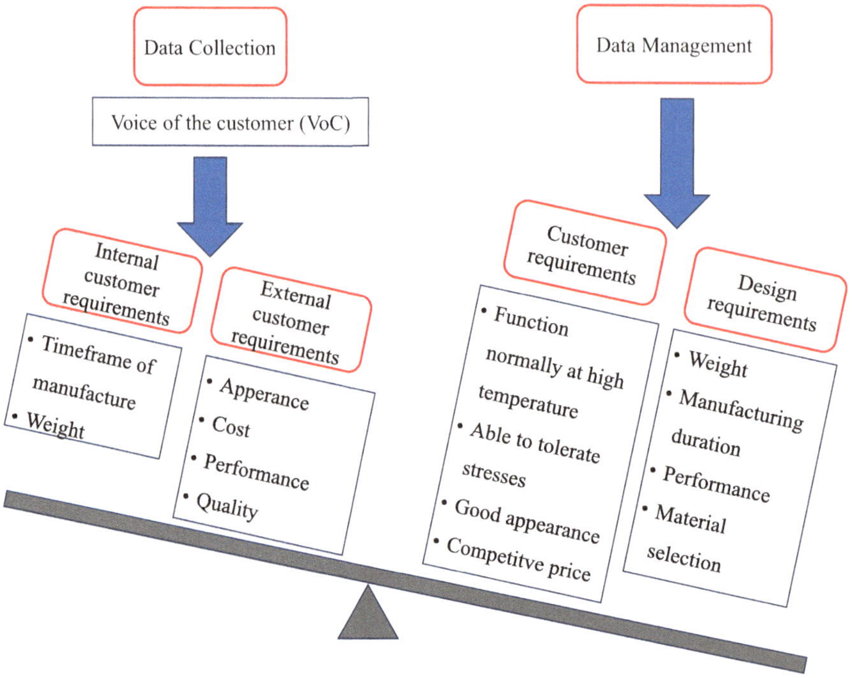

Fig. 3 Establishing the link between customer requirements and design requirements

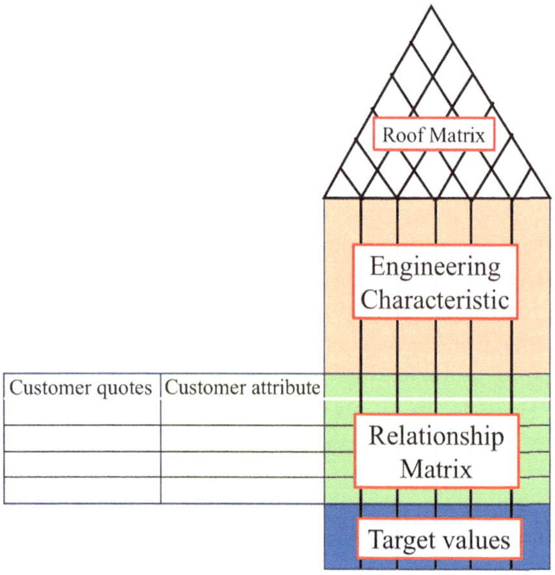

Fig. 4 The structure of the house of quality

The roof of the HoQ is the tradeoff matrix that would be crossmatched with requirements and specifications to highlight any compromises that need to be made to arrive at the best fit to customer requirements. The final step is filling the bottom row of the house with the result of the QFD. These are precise and measurable characteristics of a product that have been derived from general customer demands.

QFD in Additive Manufacturing

A direct implementation of QFD can be seen in additive manufacturing, where customer requirements are met throughout competing stages from design to product manufacturing. The different phases of QFD are linked and built on each other. Understanding these cascading connections and the 'how' of one phase feeds the 'what' of the next phase, is the secret to successfully utilizing QFD.

To implement the final QFD, the major processes include:

4.1 Preparation of the 3D Part and Selection of the Additive Manufacturing Technology

There are several types of 3D printers, each with a distinct set of materials, depending on what the consumer needs. The following are some well-known 3D printing technologies: Stereolithography (SLA), Specific Laser Sintering (SLS), and Fused Deposition Modeling (FDM), which make for quick techniques to create 3D parts. An example of the preparation and selection of the 3D printing technology is shown in Fig. 5. The first step is turning the idea into a CAD file, and then depending on the desired properties the type of technology will be selected. Then the properties of the 3D printing material and technology used dictate two major tracks: metal and plastic. For the case of a metal

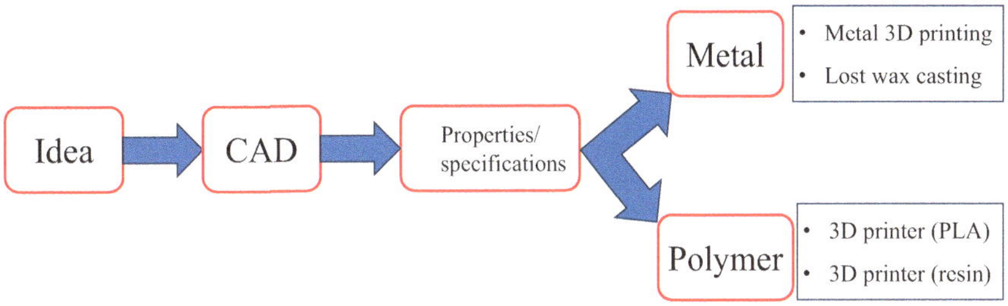

Fig. 5 Preparation and selection of the 3D printing technology

end-product, lost wax, which utilizes a temporary polymer 3D printed part to create a mold for metal casting (more on that in Chap. 7), or metal 3D printing can be used; however, for the case of a polymer end-product a single-print polymer 3D printed part can suffice.

4.2 Part Characterization Using Standard Testing Parameters (ASTM1161, ASTMB962)

Under this procedure, if the material obtained satisfies the required quality criteria, it is printed using the above-mentioned technologies. In the case of lost wax casting, selecting the best technique for casting products can make a big impact in terms of quality, functionality, cost-effectiveness, and other factors. A selection guide has been proposed by Pennsylvania Precision Cast Parts (PPCP) (a market leader in the production of high-quality castings) for a wide range of parts, sectors, and uses to evaluate the different processes. The implementation steps are shown in Fig. 6.

4.3 Post-Processing

Used to enhance the mechanical characteristics and surface quality of the final parts. The common methods used for

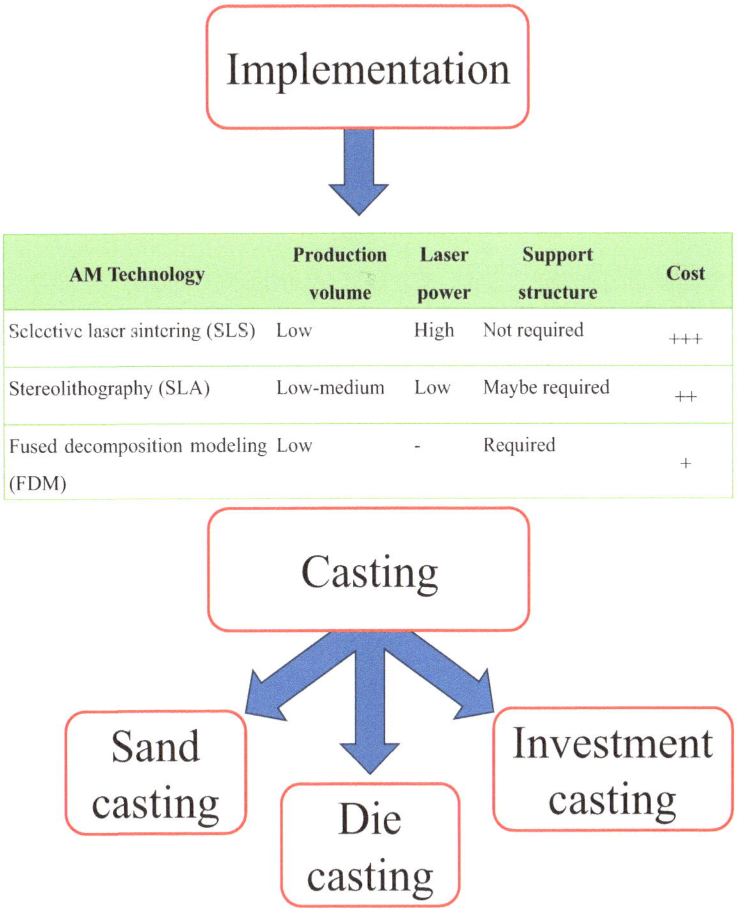

AM Technology	Production volume	Laser power	Support structure	Cost
Selective laser sintering (SLS)	Low	High	Not required	+++
Stereolithography (SLA)	Low-medium	Low	Maybe required	++
Fused decomposition modeling (FDM)	Low	-	Required	+

Fig. 6 Implementation of 3D printing technology

post-processing cast parts include heat treatments, sandblasting, anodizing, electroplating, electrophoresis, and powder coatings.

5 Life Cycle Analysis of Additive Manufacturing

Life Cycle Analysis (LCA) is a method that is used to evaluate the overall environmental and economic impacts of products and services or the impact of each phase across their life cycle. There are four main stages that are included in typical LCA, which are raw material extraction and processing, manufacturing, transportation, usage, and end-of-life, as shown in Fig. 7. Having different stages implies different LCA approaches, which are based upon the point that the assessment ends on. A cradle-to-grave LCA means that the analysis starts from material extraction and halts at the end-of-life stage, whereas cradle-to-gate, for example, ends at the usage stage. Moreover, a functional unit is an integral part of LCA, which defines the reference point upon which all calculations are based. For example, a reference prototype with a specific geometry can be used to fairly compare between different additive manufacturing technologies or different materials. There are multiple software available that can be used to conduct LCA, each with its own material library and manufacturing, transportation, and usage options. Some of these are Ansys Granta EduPack, openLCA, and Ecoinvent, which cater to their users with friendly interfaces and ease of usage.

Now, while it may seem that products printed from different materials serve similar goals, the choice of material dictates the product's impact ecologically and economically. For example, a reference PLA product's recycling phase possesses the highest impact in terms of climate change, fossil fuel depletion, human toxicity, ozone depletion, acidification, and particulate matter formation potentials, above 50% in comparison to other stages. However, when taking into account the overall impact, ABS compares much worse to PLA and PETG across all environmental impact potentials and stages, whereas PETG has the lowest (Kumar et al., 2022). This clearly shows that the choice of material plays a significant role in influencing the decisions of clients and manufacturers alike, especially with net-zero goals that are set out globally.

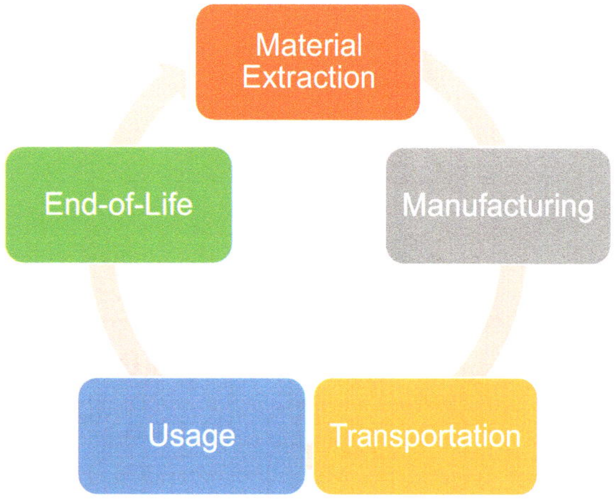

Fig. 7 Typical life cycle analysis stages

Another thing that sets additive manufacturing apart from its counterparts in a life-cycle analysis setting, is that traditional manufacturing methods often result in substantial waste due to the subtractive processes involved, whereas additive manufacturing builds objects layer by layer, using only the material necessary for the final product, which heavily influences the effect of the manufacturing stage.

Additionally, 3D printing significantly reduces production time and resource consumption. The ability to rapidly prototype and produce complex geometries without the need for specialized tooling or molds accelerates the development cycle and reduces the time to market for new products. This efficiency translates to lower energy consumption and a more sustainable manufacturing process overall.

Additive manufacturing is often classified as "green" due to its inherent efficiency and minimal waste generation. The only material waste comes from support structures, which are required for certain geometries during the printing process. These supports, however, are composed of a minimal amount of material and can often be recycled or reused which can offset negative economic and environmental effects across a product's life cycle.

References

Alparslan, A. (2020). The influence of talent management on job satisfaction and organizational commitment: A study in manufacturing sectors. In *Near East University*.

Dadić, Z., Živković, D., & Čatipović, N. (2016). Tribological wear mechanisms of molds for high pressure die casting. *Metalurgija, 55* (2), 249–252.

House of quality | Explanation with example - IONOS. (n.d.). Retrieved July 20, 2024, from https://www.ionos.com/digitalguide/online-marketing/web-analytics/house-of-quality/

Kumar, R., Sharma, H., Saran, C., Tripathy, T. S., Sangwan, K. S., & Herrmann, C. (2022). A comparative study on the life cycle assessment of a 3D printed product with PLA, ABS & PETG materials. *Procedia CIRP, 107*, 15–20. https://doi.org/10.1016/J.PROCIR.2022.04.003

Zare Mehrjerdi, Y. (2010). Quality function deployment and its extensions. *International Journal of Quality & Reliability Management, 27*(6), 616–640. https://doi.org/10.1108/02656711011054524

Additive Manufacturing and Sustainability

Abdul Hai Alami, Ahmad Yasin, Mohamad Ayoub, Shamma Alabdulla, Fabrizia Devito, Adnan Alashkar, and Ayman Mdallal

Abstract

Sustainability has surpassed being an abstract concept into being a necessary pillar of any value-adding process. Classical manufacturing processes and equipment, especially newly designed ones have had sustainability retrofitted in one form or another. With additive manufacturing, there is ample chance to incorporate sustainability into the core of the processes from the design phase, to materials selection and eventually during execution. This chapter highlights the opportunities that exist at the core of additive manufacturing to achieve unprecedented levels of sustainability. From recycling to repurposing and ending with a clear implementation of circular economy concepts, additive manufacturing ushers a new age of augmenting modern sustainability paradigms into what will become mainstream manufacturing technologies.

1 Introduction

The availability and diversity of Earth's natural resources play a crucial role in shaping how goods and services are produced and offered. Moreover, the selection and utilization of manufacturing processes determine how natural resources, raw materials, and components for products are sourced and integrated. Effective management of these resources is paramount, as it directly influences the efficiency, sustainability, and overall success of production operations. Sustainable management of production operations is critical in a dynamic and constantly evolving market. Organizations must adopt strategies that not only ensure the efficient use of resources but also minimize environmental impact. This creates the need for a comprehensive approach to production organization, which demands effective management practices. The idea of sustainability requires organizations to adopt a broader perspective, moving beyond the traditional focus on profit maximization. This shift challenges companies to consider aspects such as environmental and social, into their core strategies and operations. Sustainability principles, often referred to as sustainable development, emphasize the importance of balancing economic growth with environmental stewardship and social responsibility. Sustainable development involves adopting practices that minimize negative environmental impacts, such as the reduction of harmful gas emissions, the conservation of natural resources, and the promotion of biodiversity. This entails fostering positive social outcomes, including fair labor practices, community engagement, and equitable access to resources.

A. H. Alami (✉) · A. Yasin · M. Ayoub
Sustainable and Renewable Energy Engineering Department, University of Sharjah, Sharjah, 27272, United Arab Emirates
e-mail: aalalami@sharjah.ac.ae

A. Yasin
e-mail: ahmad.yasin@sharjah.ac.ae

M. Ayoub
e-mail: mohamad.ayoub@sharjah.ac.ae

F. Devito
Department of Mechanical, Mathematics and Management (DMMM), Polytechnical University of Bari, Bari, Italy
e-mail: fabrizia.devito@poliba.it; fabrizia.devito@iusspavia.it

F. Devito
Department of Sciences, Technologies and Society, University School for Advanced Studies Pavia - IUSS Pavia, Pavia, Italy

S. Alabdulla
Mechanical Engineering Department, American University of Sharjah, Sharjah, 26666, UAE
e-mail: g00070854@aus.edu

A. Mdallal
Sustainable Energy & Power Systems Research Centre, RISE, University of Sharjah, Sharjah, 27272, United Arab Emirates
e-mail: Ayman.Mdallal@sharjah.ac.ae

A. Alashkar
Ph.D. Program in Materials Science and Engineering, American University of Sharjah, Sharjah, 26666, United Arab Emirates
e-mail: b00028197@aus.edu

A. H. Alami (ed.), *Sustainable Additive Technologies and Innovative Manufacturing Techniques*, Advances in Science, Technology & Innovation, https://doi.org/10.1007/978-3-031-73657-5_2

Additive manufacturing (AM) transformed the industry by allowing the development of complicated models and structures with high accuracy and minimal material wastage as well as a significant reduction in material usage compared to conventional manufacturing techniques (Baumers et al., 2016; Mobarak et al., 2023). With the increasing global focus on sustainability, researchers, and industry professionals have been shifting interest to the potential of AM in supporting sustainable practices. One of the primary advantages of AM in terms of sustainability is its ability to utilize recyclable and biodegradable materials. This enables the reuse of manufacturing by-products but also promotes the creation of innovative, environmentally friendly products customized for specific uses (Puppi & Chiellini, 2020). Furthermore, the environmental advantages of additive manufacturing are further indicated by its capacity for economic sustainability. Through the implementation of localized production, AM can decrease reliance on complex supply chains and reduce the potential hazards linked to disruptions in global manufacturing. This process of localization promotes economic sustainability and has the potential to stimulate the growth of sustainable local industries, especially in regions that are working towards reducing their environmental impact while promoting their economic growth (Ford & Despeisse, 2016).

This chapter introduces the concept of sustainable management and production, manufacturing pillars and sustainable manufacturing, as well as circular economy aspects of additive manufacturing. Finally, an environmental and economic comparison between traditional manufacturing techniques and additive manufacturing is presented.

2 Sustainable Management of Production

While the pursuit of financial goals remains a critical aspect of business operations, it is increasingly being reframed within the context of sustainability. Companies are acknowledging that financial performance and sustainability are not mutually exclusive but can be synergistic. By embracing sustainable practices, organizations can cut costs through improved resource efficiency, boost their brand reputation, and reduce risks related to environmental and social challenges. Moreover, sustainable development encourages innovation and opens new market opportunities. Companies that invest in sustainable technologies and practices are better positioned to meet the growing consumer demand for eco-friendly and socially responsible products. This approach not only supports the environment and society but also drives long-term profitability and competitiveness.

The idea of sustainability demands a holistic approach to business strategy, where environmental and social values are integrated alongside financial objectives. Sustainable development represents a comprehensive framework that allows organizations to strategize based on viably sound environmental and social goals rather than focusing solely on the economic aspect. This balanced approach ensures that companies contribute to a sustainable future while achieving their business goals. More energy-efficient use at a fair price is encouraged by the emerging world economy's high competitiveness in the market. Sustainability seems like an interesting approach to overcome these difficulties while bringing about extra advantages. The Brundtland Commission of the United Nations provided its most well-known definition, which states that ``sustainable development is the development that satisfies the requirements of the present without affecting the ability of future generations to fulfill their own needs' (Pagone et al., 2016). Sustainable development has received major interest in the manufacturing sector over the past few years due to rising recognition of the social advantages of green technologies (Shahzad et al., 2022). Manufacturers play an important role in sustainable development; nevertheless, it additionally represents a major issue for companies that produce high-tech goods. Product development, services, and their transfer can all help reduce global warming while also enabling businesses to compete on a global scale (Bag et al., 2021). Sustainable management ensures that all the practices are functioning while adding environmental, social, and economic benefits. To achieve sustainable development, sustainable management through all the processes along with sustainable production must be aligned. Figure 1 illustrates the principles of production organization, emphasizing the integration of sustainability into every stage of the production process.

Manufacturers are increasingly leveraging the advanced capabilities of rapid prototyping and additive manufacturing to enhance the sustainable aspects of product manufacturing

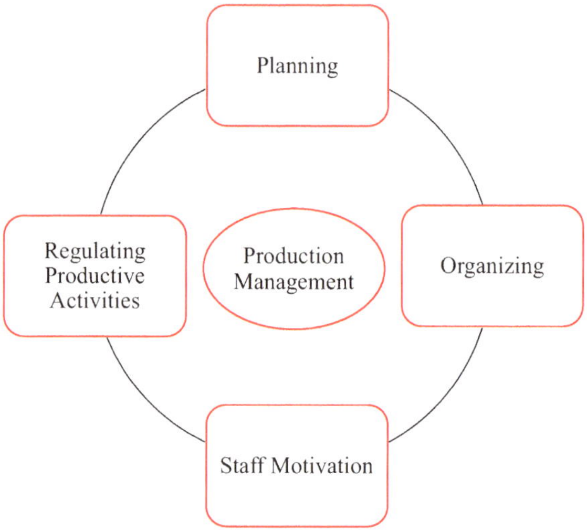

Fig. 1 Effective management influence in production processes

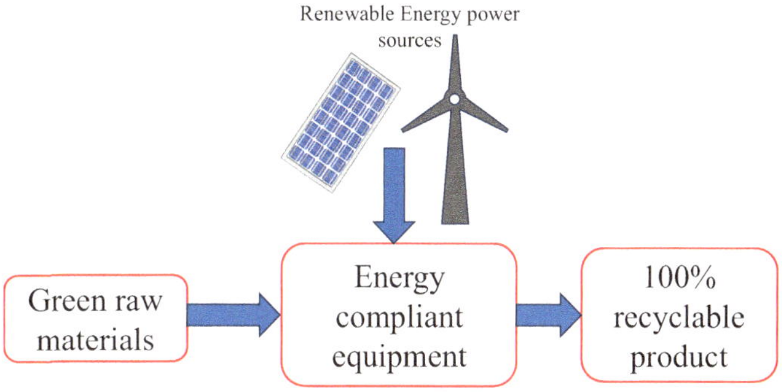

Fig. 2 Summary of manufacturing pillars

from the early stages of design and development. These technologies enable companies to create detailed prototypes quickly and efficiently, reducing the need for extensive material usage and minimizing waste. By incorporating sustainable design principles early on, businesses can avoid high costs and resource inefficiencies typically associated with traditional manufacturing methods.

Furthermore, the implementation of sustainable practices extends to later phases of the production process. Smart manufacturing systems, characterized by the use of advanced technologies such as the Internet of Things (IoT), artificial intelligence (AI), and data analytics, enable real-time monitoring and optimization of production operations. Lean manufacturing principles, which focus on minimizing waste and maximizing value, complement these smart systems by ensuring that every aspect of the production process is as efficient and sustainable as possible.

The effective management of natural resources and the adoption of innovative manufacturing technologies are critical for achieving sustainability in production operations. By integrating sustainable practices from the initial design stages through to the final production phases, organizations can enhance their operational efficiency, reduce environmental impact, and ensure long-term success in a rapidly changing market (Kazakova & Lee, 2022).

3 Sustainable Manufacturing Pillars and Challenges

The drive toward sustainability in all aspects of modern life has gathered momentum in the past decade. The focus on building efficiency and environmental consciousness in all processes and products is part of the presentability and longevity of most manufacturing establishments.

Any manufacturing process is usually energy-intensive to convert raw materials into useful products. There are doctrines to follow to dub any process sustainable, which can be broken down into each facet of the manufacturing process. For example, selecting raw materials that are largely recycled from former products is considered a strong pillar of a sustainable manufacturing ecosystem. The criteria to select suitable equipment include having good energy compliance, and this can be observed if the equipment is made from quality components and runs at good efficiency, with minimal friction, heat, or sound energy losses. This aspect can be enhanced further if the input power to the manufacturing process is from a renewable resource. Also, the design process of products should incorporate their recyclability into their lifecycle analysis, and hence sustainability grows as a culture in the manufacturing facility rather than being a burden. Figure 2 highlights a summary of these sustainable manufacturing pillars.

The sustainable manufacturing process, however, still faces several challenges that preclude its widespread adoption. The primary challenge is the inevitability of a fundamental overhaul of the production line, starting from product design to after-sales services. The transformation into green raw materials has a hidden, and hefty, price tag since replacing a conventional material such as steel with a fully recyclable material such as aluminum requires a design change and comes at a much higher monetary price. Similarly, if a novel material such as a fiber-reinforced polymer is used, not only does this impact the price, but it also deals a fatal blow to the expected recyclability. Also, seeking independence from the grid for the power-intensive manufacturing and production processes will constrain sustained productivity, which will be dependent on available sun or prevalent winds. Thus, the concept of sustainable manufacturing is a far-reaching paradigm that requires the collaborative effort of all key beneficiaries to build and sustain this ecosystem. This includes raw materials suppliers and transporters, special industrial areas with renewable energy connectivity, product design-for-sustainability, and neutral emissions as well as the amenability of the products for recyclability into new products.

4 Sustainable Manufacturing

Sustainable manufacturing refers to a manufacturing process that effectively minimizes waste and reduces its environmental impact. The primary objective of sustainable manufacturing is to achieve environmental friendliness and sustainability through the implementation of sustainable techniques and practices that influence operational concepts, product development, and process design (Rao, 2013). These practices not only reduce the consumption of raw materials and energy but also lower emissions and waste generation throughout the production lifecycle. Key techniques for achieving sustainable manufacturing include the adoption of energy-efficient technologies, waste reduction methods, and the use of environmentally friendly materials. Accordingly, manufacturers can substantially decrease their environmental footprint and improve production efficiency by integrating the aforementioned elements into their operations.

Leading manufacturers have already undertaken essential actions to bolster their contributions to social, economic, and environmental development (Ching et al., 2022). These actions involve a diverse range of strategies, including enhancing supply chain transparency, promoting ethical labor practices, investing in renewable energy sources, and advancing corporate social responsibility initiatives.

Manufacturers are utilizing the advanced benefits of rapid prototyping and AM to incorporate sustainability into product design throughout the initial stages of development. Rapid prototyping allows for the creation of detailed, functional models using minimal resources and a shorter time. Additive manufacturing, on the other hand, builds products layer by layer and uses only the material necessary for the final product, further minimizing waste and material costs.

By integrating sustainable practices across all phases of manufacturing, from product design and development to full-scale production, manufacturers can significantly reduce their environmental impact.

5 Circular Economy

The circular economy is a system focused on replacing end-of-life waste disposal with strategies for minimizing resource use, reusing, recycling, and recovering materials. It represents a transformative approach to production and consumption that prioritizes the sustainable management of resources (Kazakova & Lee, 2022). This approach goes beyond conventional recycling practices by emphasizing the extension of the life span of specific resources and materials. By reimagining the lifecycle of products and materials, the circular economy seeks to create closed-loop systems where waste is minimized, and the value of resources is maximized.

At its core, the circular economy is infused with the principles of sustainability, aiming to reduce environmental impact and resource consumption while simultaneously optimizing economic performance. Moreover, the circular economy challenges the linear "take-make-dispose" model of production and consumption, promoting instead a regenerative system designed to restore and retain value within the economy. Implementing a circular economy involves several key strategies listed below and also shown in Fig. 3, and are broken down as follows: (i) Resource Efficiency: Optimizing the use of raw materials and energy throughout the production process to reduce waste and lower environmental impact. (ii) Product Design and Longevity: Designing products with durability, reparability, and upgradability in mind, thereby extending their useful life and reducing the need for frequent replacements. (iii) Reuse and Remanufacturing: Encouraging the refurbishment and remanufacturing of products to return them to a like-new condition, thus reducing the demand for new resources. (iv) Recycling: Developing advanced recycling technologies to recover valuable materials from end-of-life products and reintroduce them into the production cycle. (v) Resource Recovery: Implementing processes to capture and repurpose waste streams, transforming them into valuable inputs for new products or energy sources.

In contrast to subtractive processes, such as milling, the additive and digital nature of additive manufacturing makes on-demand manufacturing of scrap metal for rebuilding or avoiding waste production. Such factors could also open new possibilities when designing and manufacturing for the circular economy (Sauerwein et al., 2019). The circular

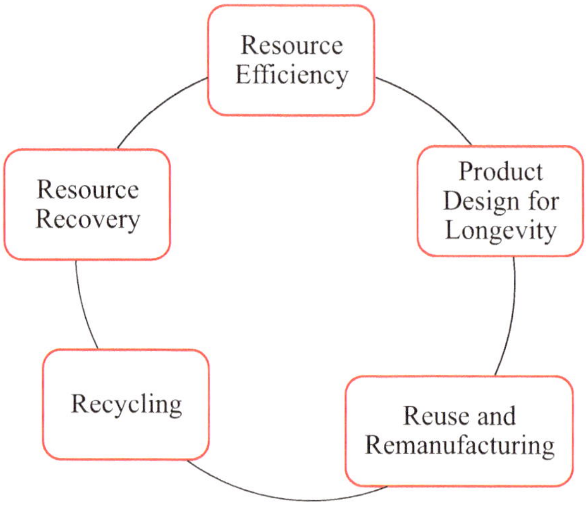

Fig. 3 Pillars of circular economy

economy, with its 3R (reduce, reuse, recycle) principles, links economic growth with environmental protection by extending the lifespan of products that have reached the end of their physical or functional utility. This approach aims to optimize their remaining value and prolong their usefulness, reducing the need for disposal (Ponis et al., 2021).

6 Environmental Impact of Conventional Manufacturing and Their Mitigation Through Sustainable Additive Manufacturing

In the product development phase, it's crucial to consider various factors, including the product's purpose, design, and disposal options like end-of-life recycling and reuse. A major focus should be on the product's environmental impact throughout its entire lifecycle, from manufacturing to disposal. This includes evaluating the technology used and the manufacturing process itself, with a particular emphasis on climate change. Key environmental concerns can be assessed by measuring the annual carbon dioxide equivalents (CO_2eq) emissions, typically expressed in tons. This measure is essential for managing emissions and assessing the manufacturing process's carbon footprint. Additionally, the effects on water and soil quality are crucial during the evaluation process.

Conventionally, the manufacturing process was mainly evaluated by four main manufacturing attributes, namely cost, time, quality, and flexibility. However, with technological advancements, additional attributes are added to the equation such as energy and resources efficiency, and waste management. These attributes are a key foundation of sustainability that needs to be considered when conducting sustainability analysis. Considering these factors raises the question of how conventional and additive technologies compare in terms of sustainability. To answer this, three principal aspects are used to indicate the environmental impact of a manufacturing process, such as resource efficiency and consumption, pollution mitigation, and waste management.

In terms of resource efficiency and consumption, additive manufacturing technologies are estimated to be up to 97% more efficient in terms of material use contrasted with conventional manufacturing methods (Peng et al., 2018). Moreover, Material usage includes not only primary materials like polymer filaments but also secondary materials such as protective gases like nitrogen or argon, and cooling water. In contrast, conventional manufacturing entails a wider usage of materials, including coolants, lubricants, and various primary materials and tooling components.

On the other hand, pollution mitigation and waste management are mainly related to the materials used and waste in the manufacturing process as well as how energy intensive the manufacturing process is. Conventional manufacturing methods, such as machining, casting, and forming, are fundamental to industrial production but often come with significant environmental drawbacks. These methods typically involve high energy consumption, extensive material waste, and substantial greenhouse gas (GHG) emissions. These processes consume a considerable amount of electricity and, when sourced from fossil fuels, increase the carbon footprint of manufacturing. Furthermore, conventional subtractive manufacturing methods, where the material is removed from a larger block, can result in significant material waste. This not only depletes resources but also adds to the overall environmental impact through increased waste disposal requirements. In addition to energy use and material waste, conventional manufacturing contributes to significant GHG emissions. The steel industry, for example, is responsible for approximately 7% to 8% of global CO_2 emissions.

Nevertheless, additive manufacturing processes can vary significantly in their energy requirements and emissions. Generally, AM technologies are known for their potential to reduce GHG emissions compared to conventional manufacturing processes. However, the energy source used to power AM machines affects the overall emissions profile. Despite the energy source used during the manufacturing process, additive manufacturing is still considered a cleaner option compared to conventional machining, with a potential reduction of 70% (Direct additive laser manufacturing vs conventional machining) in its environmental impact. Furthermore, by integrating AM technologies into the manufacturing process, the total waste produced can be reduced by 90% compared to subtractive manufacturing methods (Peng et al., 2018). Additionally, AM technologies generally use fewer non-toxic materials and chemicals than conventional manufacturing methods. For instance, common materials for 3D printing, such as Poly Lactic Acid (PLA) and Acrylonitrile Butadiene Styrene (ABS), result in an easier recycling path, which positively impacts AM's environmental footprint.

7 Economic Analysis of Conventional Manufacturing and Advantages of Additive Manufacturing Routes

In order to comprehend the economic impact of currently available manufacturing routes, conventional and additive, it is essential to dictate what limits each technology as well as its customers. Geometry is considered a limitation in conventional manufacturing, where designs with fine details or internal features (internal joints), may present a lot of difficulties for techniques such as machining and casting.

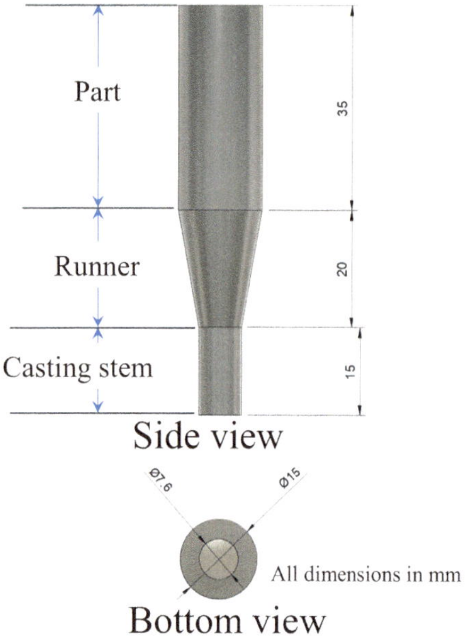

Fig. 4 Standard object for comparison–Retrieved from (Alami et al., 2023) (open access)

Whereas in additive manufacturing, geometry poses little to no limitation. Moreover, it could require more than one technique going through the conventional route of manufacturing, unlike additive manufacturing which is usually a one-step process. In terms of profitability, additive manufacturing capitalizes on independent orders from different customers, where the objective is to suit various mass customization requirements. On the other hand, conventional manufacturing benefits from mass orders and makes a profit on large batches (Additive Manufacturing VS Traditional Manufacturing: Cost and Advantages of 3D Printing Technology, n.d.). Comparing the price and energy consumption related to producing the same object, shown in Fig. 4 through CNC and the lost PLA method (a method heavily dependent on additive manufacturing) yields results that are biased towards the latter. The normalized cost (per gram) of the produced piece was around \$0.067/g produced via the lost wax method (with less kerf losses) and \$0.096/g produced via CNC material removal. Finally, the electrical energy consumption was 12.124 kWh and 17.297 kWh for the lost wax method and CNC, respectively (Alami et al., 2023).

The cost structure of manufacturing technologies, just like any venture, can be broken into fixed and variable costs. Taking different conventional manufacturing techniques, for example, capital costs are usually related to the machinery that is used for each route. Metal injection molding tooling can reach up to \$125k (Smith Metal Products | MIM Details: Tooling Cost, n.d.) in upfront costs, and CNC machining striding similar cost brackets with maintenance and mill-bits and raw materials required as variable costs. Alternatively, AM necessitates a one-time purchase of an apparatus that requires a lot less maintenance and waste of material feedstock. Additionally, lead times of traditional manufacturing techniques are much longer than those for additive manufacturing, with metal injection molding reaching up to 6 months, compared to days or weeks (as a worst-case scenario) for additive manufacturing (Metal 3D Printing Cost Analysis vs Traditional Manufacturing | Explore 3D Metal Printing Costs Compared to Traditional Manufacturing—3DEO, n.d.).

The impact of AM on the industry can be broken down into market shares, technological and design software effects, and their advancements. The additive manufacturing market size skimmed over the \$20 billion mark in 2023 and is expected to have a compound annual growth rate of 23.3% from 2023 to 2030. In just 2021, about 2 million 3D printers were shipped globally, with expectations to reach almost 21.5 million by 2030. Industrial 3D printers have dominated the market according to 2022 metrics, with a market share of 76%, and applications in automotive, electronics, and aerospace sectors. In terms of technology dominance, stereolithography has had its fair share of the market in 2022, claiming 10% of the global revenue. Expectedly, design software, which goes hand in hand with additive manufacturing technologies accounted for 36% of revenue shares of the additive manufacturing market (Additive Manufacturing Market Size Report, 2030, n.d.).

References

Additive Manufacturing Market Size Report, 2030. (n.d.). Retrieved July 14, 2024, from https://www.grandviewresearch.com/industry-analysis/additive-manufacturing-market

Additive manufacturing VS traditional manufacturing: Cost and advantages of 3D printing technology. (n.d.). Retrieved July 14, 2024, from https://www.roboze.com/en/resources/additive-manufacturing-vs-traditional-manufacturing-cost-and-advantages-of-3d-printing-technology.html

Alami, A., Ayoub, M., Yasin, A., Alashkar, A., Mdallal, A., Khuri, S., Al Abdulla, S., Aljaghoub, H., & Alshemsi, F. (2023). Lost wax casting. *Resourceedings*, *3*(3), 10–16. https://doi.org/10.21625/resourceedings.v3i3.1023

Bag, S., Gupta, S., & Kumar, S. (2021). Industry 4.0 adoption and 10R advance manufacturing capabilities for sustainable development. *International Journal of Production Economics*, *231*, 107844. https://doi.org/10.1016/j.ijpe.2020.107844

Baumers, M., Dickens, P., Tuck, C., & Hague, R. (2016). The cost of additive manufacturing: Machine productivity, economies of scale and technology-push. *Technological Forecasting and Social Change*, *102*, 193–201. https://doi.org/10.1016/j.techfore.2015.02.015

Ching, N. T., Ghobakhloo, M., Iranmanesh, M., Maroufkhani, P., & Asadi, S. (2022). Industry 4.0 applications for sustainable manufacturing: A systematic literature review and a roadmap to sustainable development. *Journal of Cleaner Production, 334*, 130133. https://doi.org/10.1016/j.jclepro.2021.130133

Ford, S., & Despeisse, M. (2016). Additive manufacturing and sustainability: An exploratory study of the advantages and challenges. *Journal of Cleaner Production, 137*, 1573–1587. https://doi.org/10.1016/j.jclepro.2016.04.150

Kazakova, E., & Lee, J. (2022). Sustainable manufacturing for a circular economy. *Sustainability (Switzerland), 14*(24), 17010. https://doi.org/10.3390/su142417010

Metal 3D printing cost analysis versus traditional manufacturing | explore 3D metal printing costs compared to traditional manufacturing—3DEO. (n.d.). Retrieved July 14, 2024, from https://www.3deo.co/strategy/costs-of-traditional-vs-additive-manufacturing/

Mobarak, M. H., Islam, Md. A., Hossain, N., Al Mahmud, Md. Z., Rayhan, Md. T., Nishi, N. J., & Chowdhury, M. A. (2023). Recent advances of additive manufacturing in implant fabrication–A review. *Applied Surface Science Advances, 18*, 100462. https://doi.org/10.1016/j.apsadv.2023.100462

Pagone, E., Jolly, M., & Salonitis, K. (2016). The development of a tool to promote sustainability in casting processes. *Procedia CIRP*, 53–58. https://doi.org/10.1016/j.procir.2016.09.001

Peng, T., Kellens, K., Tang, R., Chen, C., & Chen, G. (2018). Sustainability of additive manufacturing: An overview on its energy demand and environmental impact. *Additive Manufacturing, 21*, 694–704. https://doi.org/10.1016/j.addma.2018.04.022

Ponis, S., Aretoulaki, E., Maroutas, T. N., Plakas, G., & Dimogiorgi, K. (2021). A systematic literature review on additive manufacturing in the context of circular economy. *Sustainability (Switzerland), 13* (11), 6007. https://doi.org/10.3390/su13116007

Puppi, D., & Chiellini, F. (2020). Biodegradable polymers for biomedical additive manufacturing. *Applied Materials Today, 20*, 100700. https://doi.org/10.1016/j.apmt.2020.100700

Rao, P. N. (2013). Sustainable manufacturing—principles, applications and directions. In *28 National convention of production engineers & national seminar on "Advancements in production and operations management" F.W.*, 28.

Sauerwein, M., Doubrovski, E., Balkenende, R., & Bakker, C. (2019). Exploring the potential of additive manufacturing for product design in a circular economy. *Journal of Cleaner Production, 226*, 1138–1149. https://doi.org/10.1016/j.jclepro.2019.04.108

Shahzad, M., Qu, Y., Rehman, S. U., & Zafar, A. U. (2022). Adoption of green innovation technology to accelerate sustainable development among manufacturing industry. *Journal of Innovation and Knowledge, 7*(4), 100231. https://doi.org/10.1016/j.jik.2022.100231

Smith metal products | MIM details: Tooling cost. (n.d.). Retrieved July 14, 2024, from https://smithmetals.com/mim-details/mim-costs/tooling-cost/

Computer-Aided Design Role in Additive Manufacturing

Abdul Hai Alami, Ahmad Yasin, Adnan Alashkar, Mohamad Ayoub, Ayman Mdallal, Fatima Alshamsi, Siren Khuri, Fabrizia Devito, Shamma Alabdulla, and Haya Aljaghoub

Abstract

This chapter highlights the importance and procedure of digitizing the design idea. After a product is converted from requirements and constraints into dimensions, contours and materials, the product takes shape and moves into the tangible world. There are many benefits of converting a hand sketch into a digital file, including easier modification, sharing, and replication. The chapter concludes with a generic tutorial on basic functions found in almost all current 3D modeling software and how to export the file into formats that will be used for subsequent processing and manufacturing steps.

A. H. Alami (✉) · A. Yasin · M. Ayoub · F. Alshamsi
Sustainable and Renewable Energy Engineering Department, University of Sharjah, Sharjah, 27272, United Arab Emirates
e-mail: aalalami@sharjah.ac.ae

A. Yasin
e-mail: ahmad.yasin@sharjah.ac.ae

M. Ayoub
e-mail: mohamad.ayoub@sharjah.ac.ae

F. Alshamsi
e-mail: u20200968@sharjah.ac.ae

A. Alashkar
Ph.D. Program in Materials Science and Engineering, American University of Sharjah, Sharjah, 26666, United Arab Emirates
e-mail: b00028197@aus.edu

A. Mdallal · S. Khuri
Sustainable Energy & Power Systems Research Centre, RISE, University of Sharjah, Sharjah, 27272, United Arab Emirates
e-mail: Ayman.Mdallal@sharjah.ac.ae

S. Khuri
e-mail: SKhuri@sharjah.ac.ae

F. Devito
Department of Mechanical, Mathematics and Management (DMMM), Polytechnical University of Bari, Bari, Italy
e-mail: fabrizia.devito@poliba.it; ; fabrizia.devito@iusspavia.it

S. Alabdulla
Mechanical Engineering Department, American University of Sharjah, Sharjah, 26666, UAE
e-mail: g00070854@aus.edu

H. Aljaghoub
Management Science and Engineering Department, Khalifa University, Abu Dhabi, 127788, UAE
e-mail: 100065161@ku.ac.ae

1 Introduction

Engineered products were historically designed and built by artisans who inherited a set of skills and tools from one generation to another. The design and manufacturing processes had a blurred border separating them, as intuition and experience played major roles in product development. During the Industrial Revolution, when things had to be made in large quantities, accurately, and with high quality, relying solely on the talent of elite masters became a serious bottleneck. Another important factor is the conundrum of saving the production recipe of stages and steps. This usually relied on the rare chance of employing a talented and equally honest apprentice who would absorb the design and manufacturing ways of their master to maintain the continuity of uniqueness and quality of the products for generations to come. This has proven a sizable challenge when the need for mass production took center stage and automation rose as the new status quo.

Standardized shop drawings were a consequence of the inevitable separation of the design, manufacture, and production processes. Designers began documenting the shape, dimensions and tolerances in sketches that used drafting tools such as T-square, triangles, compasses and protractors, as shown in Fig. 1. Details of surface finish, tolerances, suggested fasteners and other manufacturing details were also documented on the drawings, which were filed and stored for later use and updated with further design iterations. The separation of design and manufacture introduced its unique drawbacks, such as instances of designing and

A. H. Alami (ed.), *Sustainable Additive Technologies and Innovative Manufacturing Techniques*, Advances in Science, Technology & Innovation, https://doi.org/10.1007/978-3-031-73657-5_3

Fig. 1 The collaborative effort of using drafting tools such as T-squares and triangles with multiple designers and drafters working on the same project (How People Worked Before AutoCAD, n.d.) https://www.wmeglobal.com/how-people-worked-before-autocad/

drafting an unmanufacturable part, that necessitated concepts such as concurrent engineering, design-for-manufacturing, design-for-assembly, and the comprehensive umbrella of computer-integrated manufacturing (CIM).

The digitalization of concepts in the form of computer-aided design (CAD) files has not only replaced huge tiltable drafting tables (the author's back still hurts) but also replaced drafting tools that were meticulously maneuvered around the drawing space to avoid smudging the fine pencil lines that hermetically intersect to form the desired shape. Various software packages are currently available for drafting and modeling tasks and also for streamlined post-processing of the designed part. Since computerized machines are available for additive manufacturing (e.g. 3D printing) or material removal processes (milling, turning, etc.) the digital data built in a 3D modeling environment allow the manipulation of said data to feed production machines as well as modeling software. Thus apart from allowing virtual collaboration on the design of products, CAD software allows streamlined dissemination of designs, straightforward design modifications, adaptation to various production equipment, material utilization and production time estimation. The data stream accompanying these processes is intensive and also very useful for quality assurance and future development of the product and process. It also allows simulation and modeling to predict the part strength and response to dynamic loading.

2 From Sketch to CAD File: The Digitization of Ideas

At this stage, the design idea of the product should be clear with its physical and functional constraints. This allows the designer to "box" the design within 3D constraints that bound the product within a certain physical size. An example is shown in Fig. 2. This back-of-the-envelope approach is a good starting point prior to delving into full digitization by establishing a cornerstone for any further additions or modifications.

There are many CAD software available in the market, and many are available with a free license for educational institutes. Some software packages are dedicated to 2D drawings, such as building, ducting or wiring layouts (e.g. AutoCAD® or PunchCAD®) while others offer three-dimensional modeling (e.g. SolidWorks® or Fusion360®). The 3D modeling software are more important within the context of this book as they allow parts to be designed, analyzed and then tessellated for processing within 3D printing software.

Contrary to machining processes, additive manufacturing processes require the addition of supports whenever there is an overhand in the design. This may not be intuitive for the novice 3D printer user, but it is an important aspect of the part design process. Neglecting to add supports would result in a filament spaghetti as shown in Fig. 3b while attempting

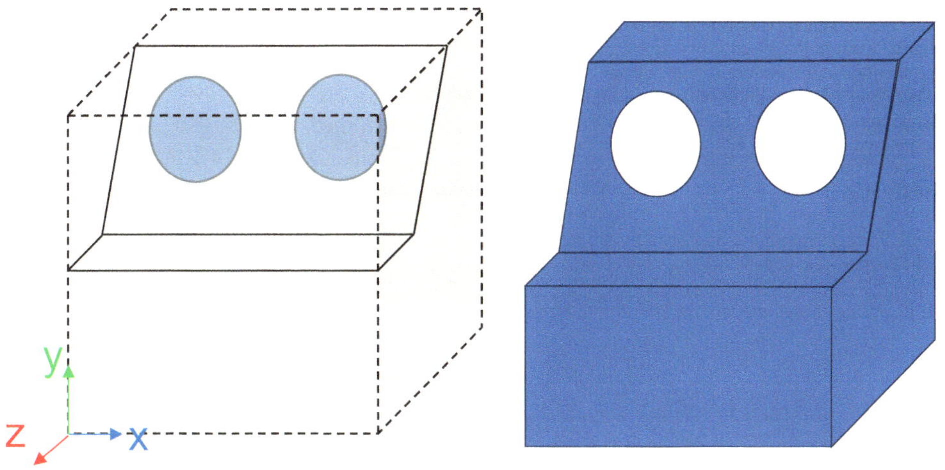

Fig. 2 Box-in approach to design commencement

to print the overhang of the part shown in Fig. 3a. Adding supports will increase the build time and also the material consumption, but the build will not be complete without them. The part shown in Fig. 3 can even be printed with a different orientation as shown in Fig. 3d, which eliminates the need for supports altogether. Thus, a designer should experiment with various orientations until the one that requires less (or no) support should be adapted. This approach is a design-for-manufacturing one, which facilitates the additive manufacturing of the designed product and should be considered while planning for part modeling.

3 Basic Functions of 3D Modeling Software

Nowadays, most 3D modeling software adopts a consistently comparable approach to generating 3D objects. Packages such as SolidEdge®, ProEngineer® that became Creo®, SolidWorks® and Fusion360®, are all based on similar engines to generate 3D objects. Thus, learning one of these software packages would pave the way for applying the same knowledge to other packages as well. The Internet is full of detailed tutorials and tips on how to start from a novice level till one reaches expert status, especially when consulting the website of the company that develops the software.

This section will focus on using Fusion360® from AutoDesk®. The software has a free license for education and is straightforward to learn and powerful enough for additive manufacturing purposes. The chosen part design will highlight the basic functions within the software, which is available in other commercial packages. More advanced and sophisticated features will be left to the reader to explore but will be highlighted for general benefit. For example, the software has a dedicated Finite Element structural analysis feature that can help determine the strength of the built part. Another feature is the direct 3D printing tab that allows the software to prepare the 3D file for specific 3D printer brands. These features are important but not essential if one is satisfied with software that comes bundled with a particular 3D printer.

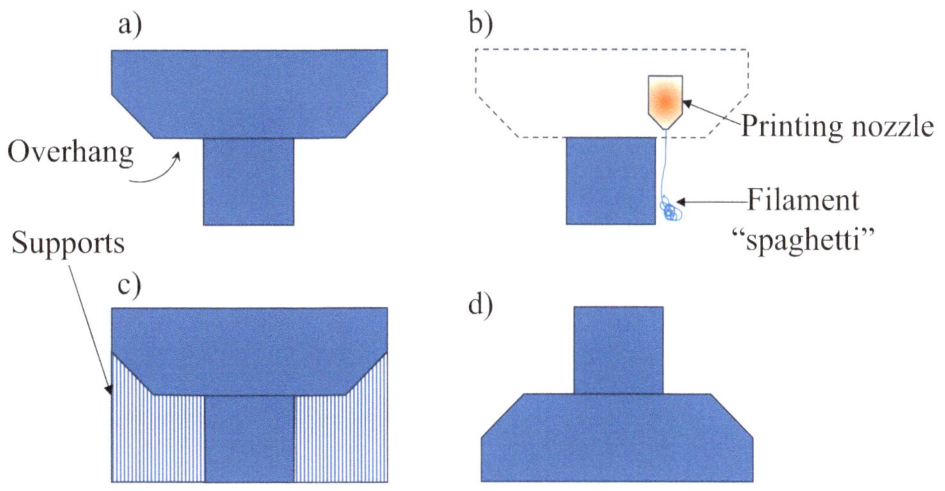

Fig. 3 **a** A part designed with an overhand will be impossible to print directly as **b** the printing nozzle will start depositing filaments in thin air. It is important to **c** incorporate disposable supports that will assist in building overhangs and then later st

4 The Common Start

Apart from ready-made objects (spirals, spheres, etc.) that can be dropped in the workspace then sized and/or oriented according to the user design, almost all commands within 3D software would begin with a 2D sketch. In Fusion360®, the user gets to select the 2D plane (x–y, x–z or y–z) that is most appropriate for the design orientation as shown in Fig. 4. This, however, has little subsequent importance as the 3D shape can be re-oriented (rotated and translated) based on eventual presentation requirements.

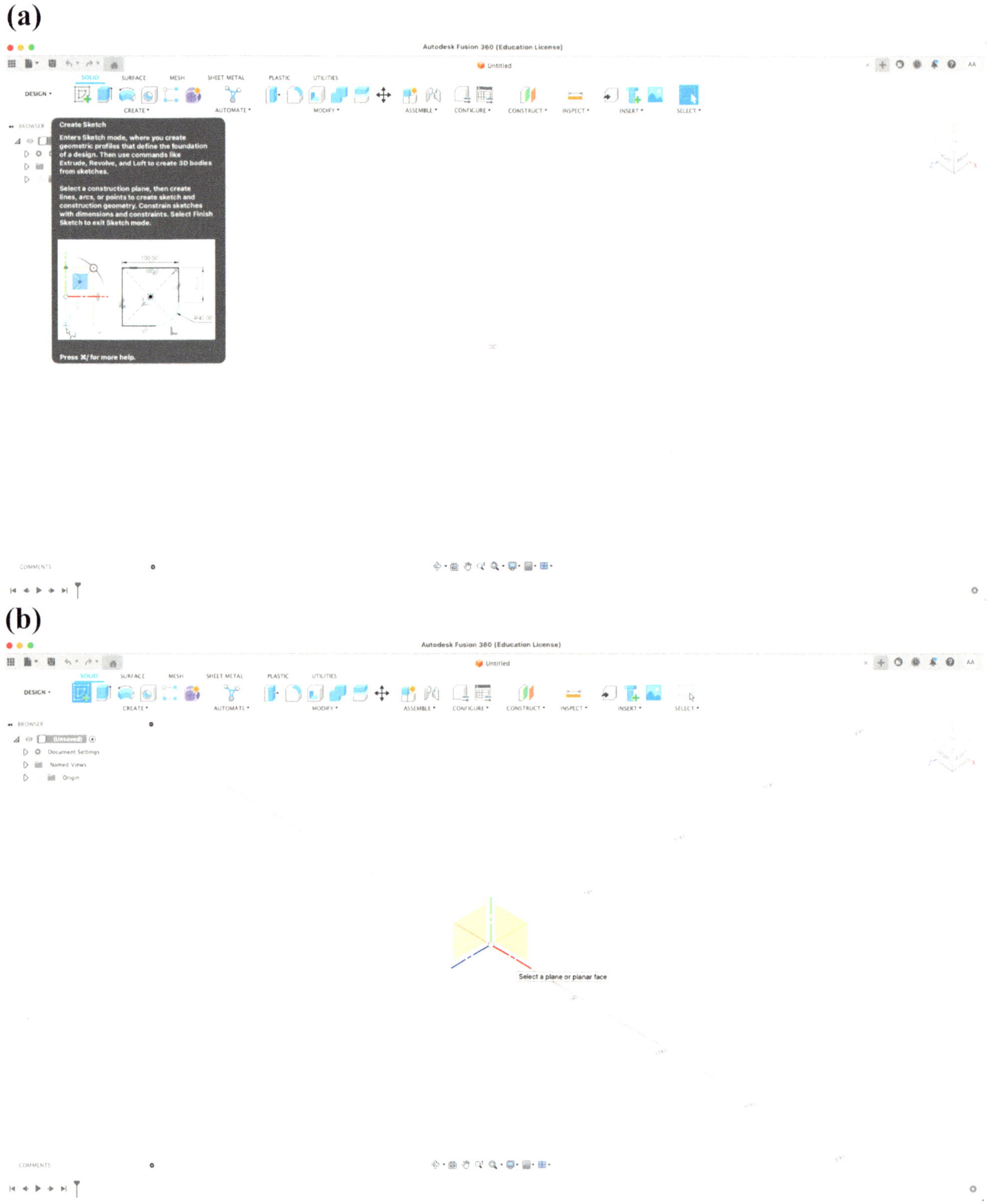

Fig. 4 **a** Selecting the 2D sketch command and **b** selecting the working plane in Fusion360®

5 Extrude

The extrude command is one of the most used commands in 3D modeling and refers to a solid object that results from extruding a 2D sketch in a certain direction for a certain distance. The path could be straight, resulting in right shapes or converging/diverging, resulting in tapered shapes depending on the prespecified taper angle.

To produce the object shown in Fig. 2, an extrusion operation can be applied along the negative x-axis to a 2D sketch on the y–z axis. The reader is encouraged to give this some thought and think of better modeling options that allow arriving at the desired shape more straightforward than the suggestion above.

The steps are detailed in succession in Fig. 5. After the y–z plane is selected, it takes the user to the sketch window shown in Fig. 5a. The "create" part of the menu contains the basic line, arc and circle creation commands. For example, when the line command is pressed, the mouse cursor shape changes and the user should now select the starting point of this line, as shown in Fig. 5b. The length and angle of this line are two options to be chosen next and can be done either by holding and dragging the mouse or by inputting these values in the boxes that appear along the drawn line as seen in Fig. 5c.

Once the shape is "closed" the enclosed profile turns a hue of blue as shown in Fig. 5d. This is important to check at this stage as subsequent commands cannot be executed if the end points of the profile do not snap.

Once the sketch is done and the "Finish Sketch" button is pressed either from the contextual menu or by pressing the green button shown in Fig. 5d, the 2D profile is now ready to be processed into a 3D object as shown in Fig. 6a.

The fully extruded shape is shown in Fig. 6b. There are several options on how the software should perceive the shape: a new component, a joint object, a cut object or a new body. It is wise to keep the option to edit the part or translate it with respect to other parts by saving it as either a new body or a new component.

The circular holes can be added in a similar manner and extruded but with the *cut* option. The construction plane in this case is the inclined surface and extruded into the body to the design extent. The shape finally looks as in Fig. 7a and b.

6 Revolve

Another versatile command that takes advantage of symmetry around an axis, or axisymmetric bodies. Just like a potter shapes the clay on a pottery wheel, the rotation of the platform and the even pressure from the hands produces a solid object that is symmetric around the axis of rotation. In machining, a lathe operates on the same principle.

Creating a revolved object starts the same as with extruded objects, with selecting the plane to draw the profile to be extruded as shown in Fig. 8.

After the profile is drawn and locked (enclosed area turns into a light blue shade), there is one last step before pressing "Finish Sketch": the axis of rotation has to be specified.

It is good practice to construct dedicated planes or axes, and this is accessed from the "Construct" portion of the menu. As seen in Fig. 9a, there are several options to construct the axis from, including the one that will be used here, which is "Axis Through Two Points". This step may not be necessary as the axis of rotation can be specified as the two vertices of the profile in Fig. 8, but creating a separate axis grants more control over the revolve operation. The axis is built and shown in Fig. 9b.

There are several options that can be invoked when performing the Revolve command, such as the sweep angle and direction, as shown in Fig. 10a. Once the profile is selected along with the axis, the command is complete and the object appears as seen in Fig. 10b.

If there was an offset between the profile and axis of rotation, then the shape will have a hole at the center and will look like the object in Fig. 11a. Note that there are perforations at the outer surface of the object that were constructed using the Pattern command, which is another very useful command. Patterns can be either circular or rectangular and are constructed by drawing one feature only and then invoking the Pattern command as shown in Fig. 11b. The number and distribution of the features can be controlled using the command, which speeds up the design process.

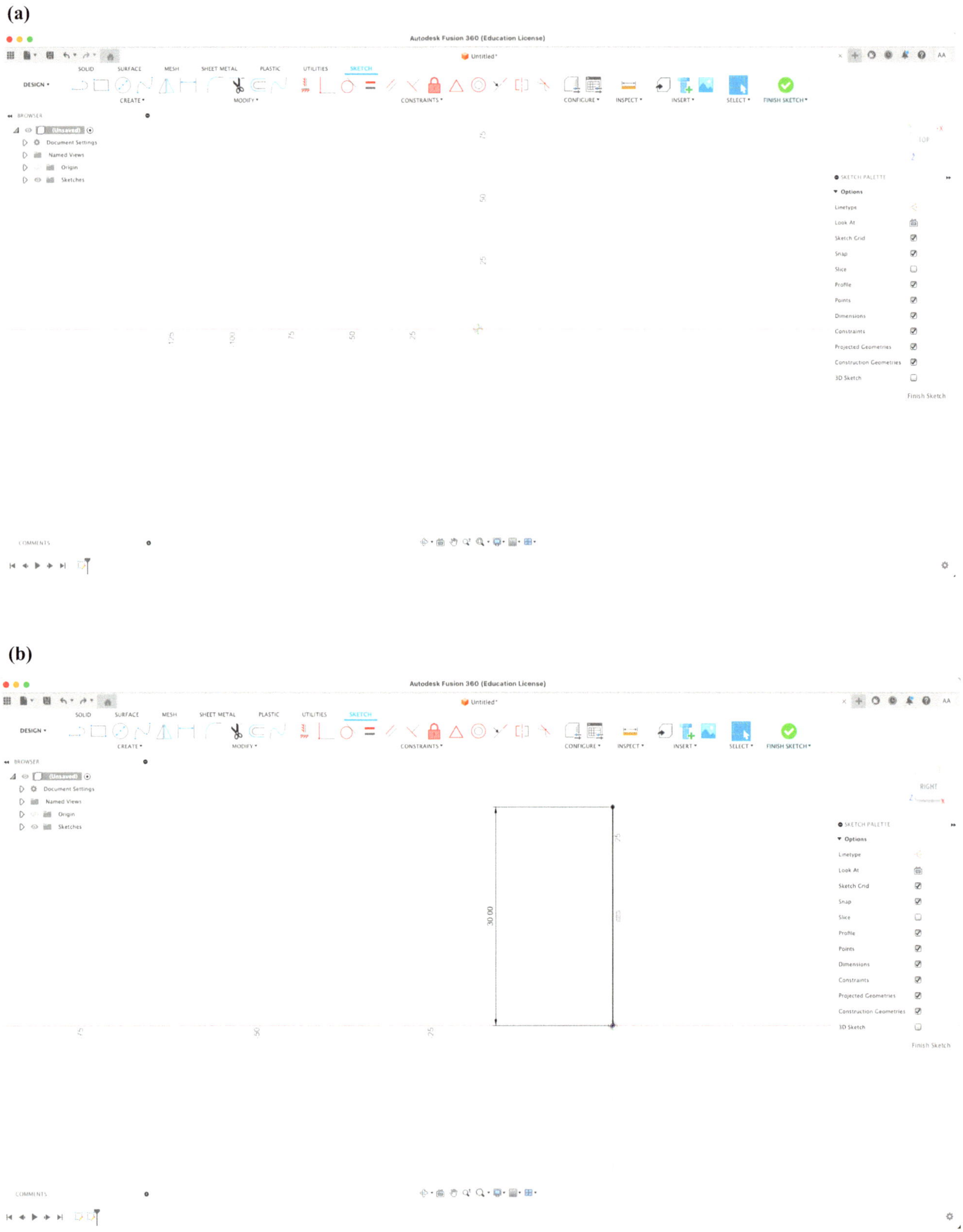

Fig. 5 **a** Entering the sketch screen in Fusion360©, **b** drawing a line, **c** angles and **d** ensuring that the sketch is closed (with the internal light blue hue visible)

(c)

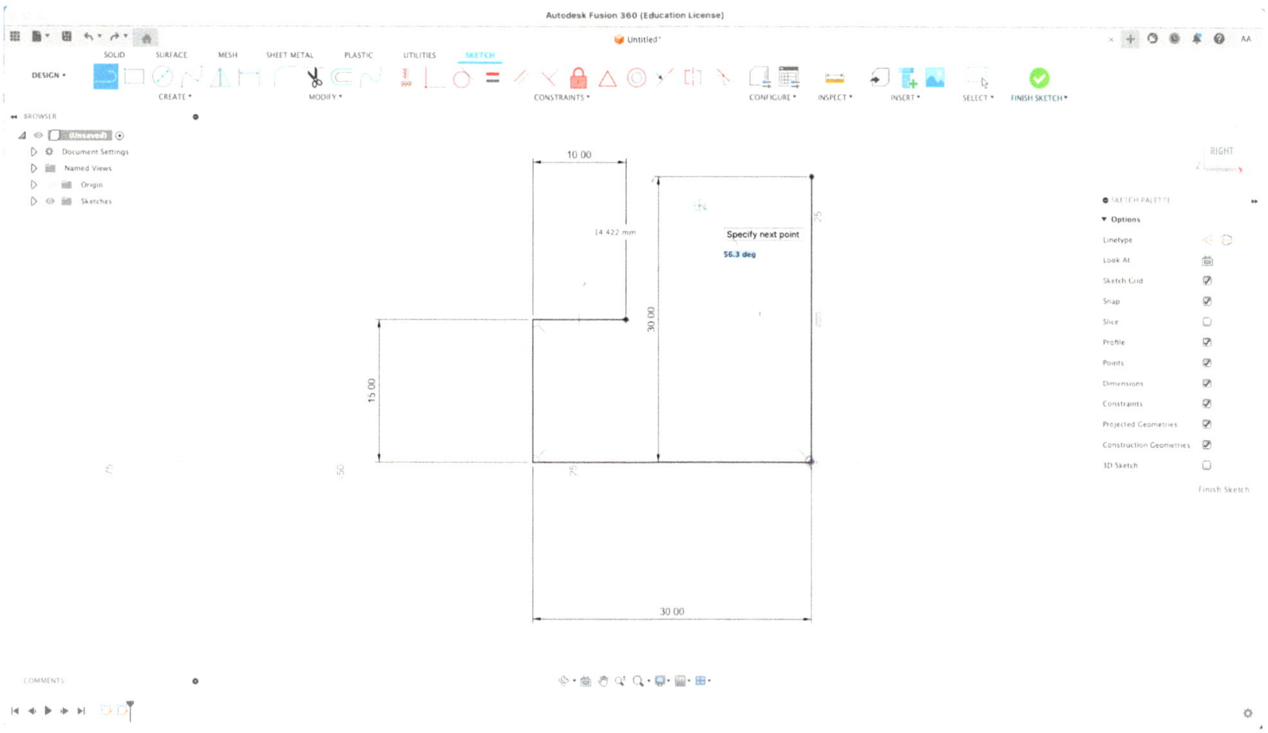

(d)

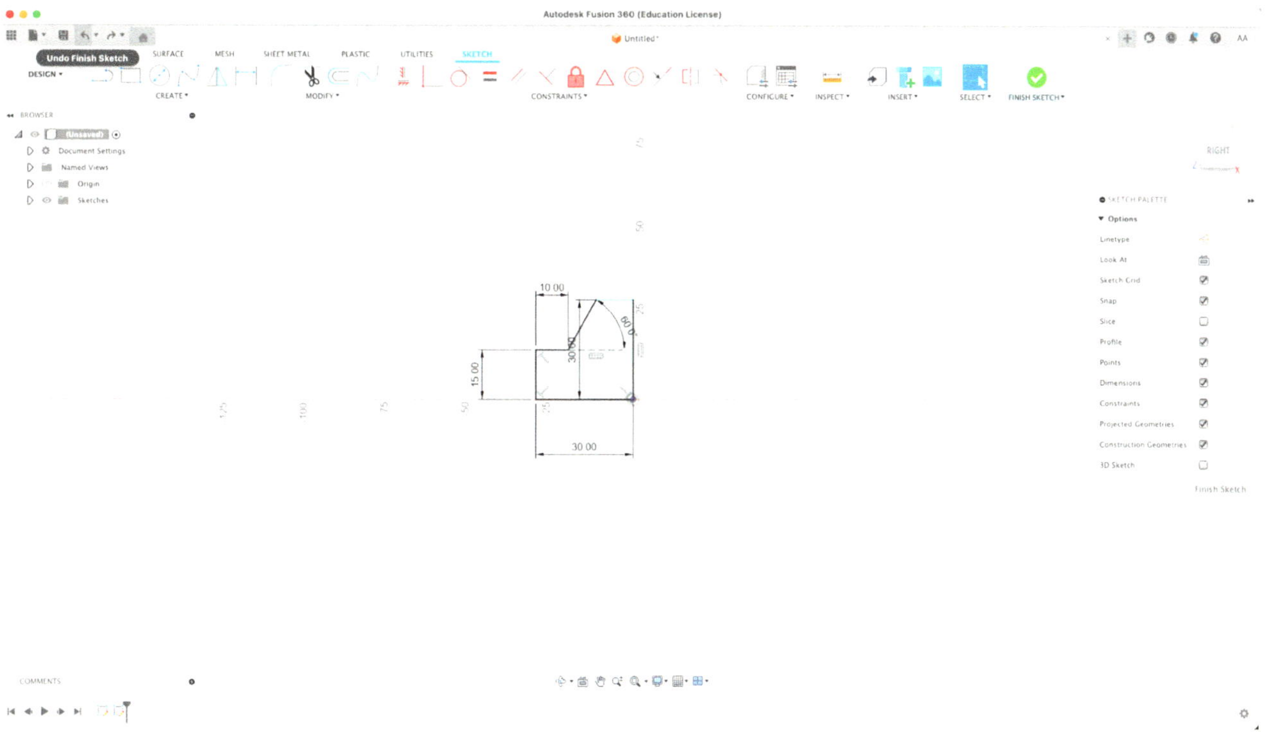

Fig. 5 (continued)

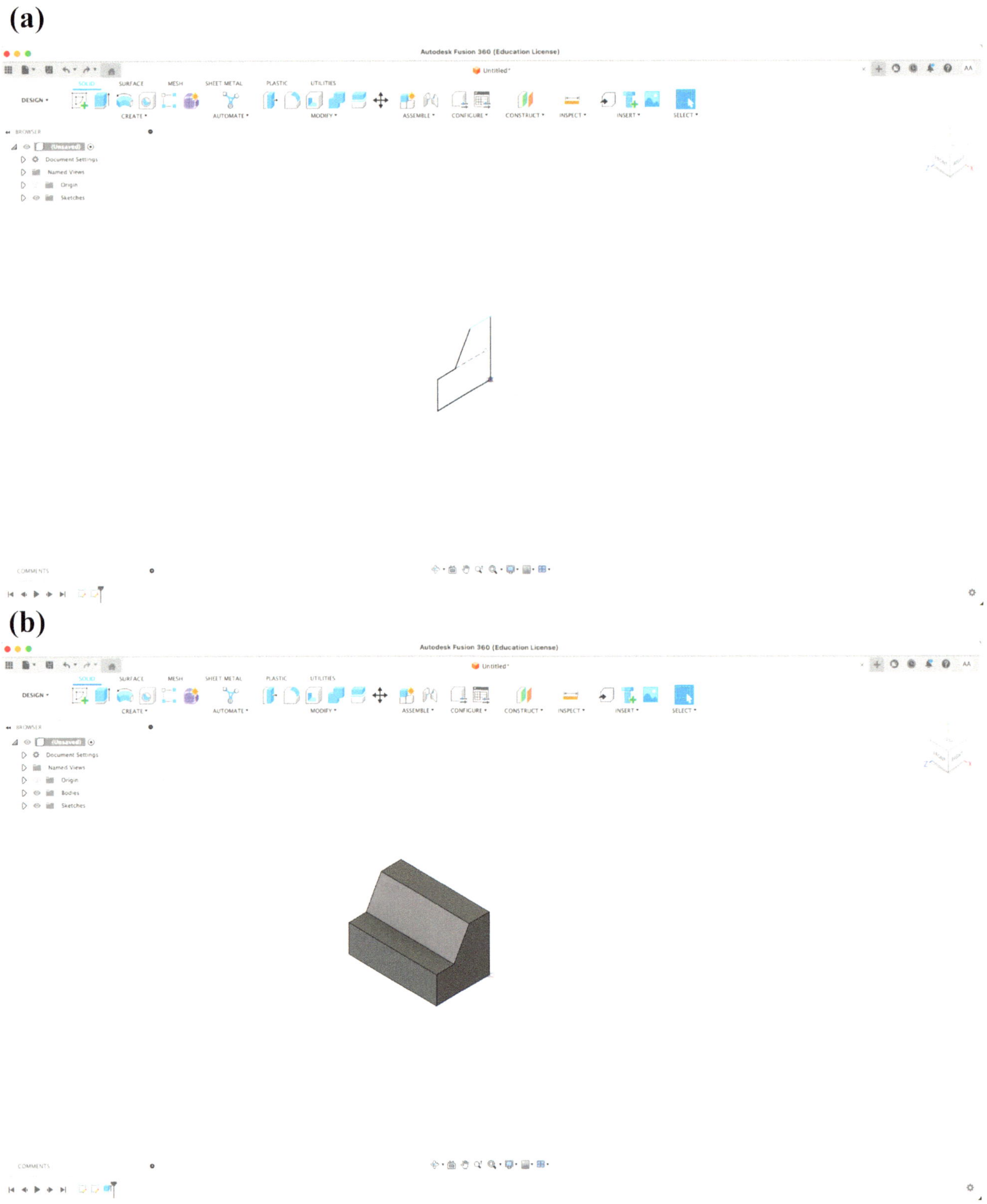

Fig. 6 **a** After the sketch is done and closed by clicking (Finish Sketch) the **b** Extrude command is applied

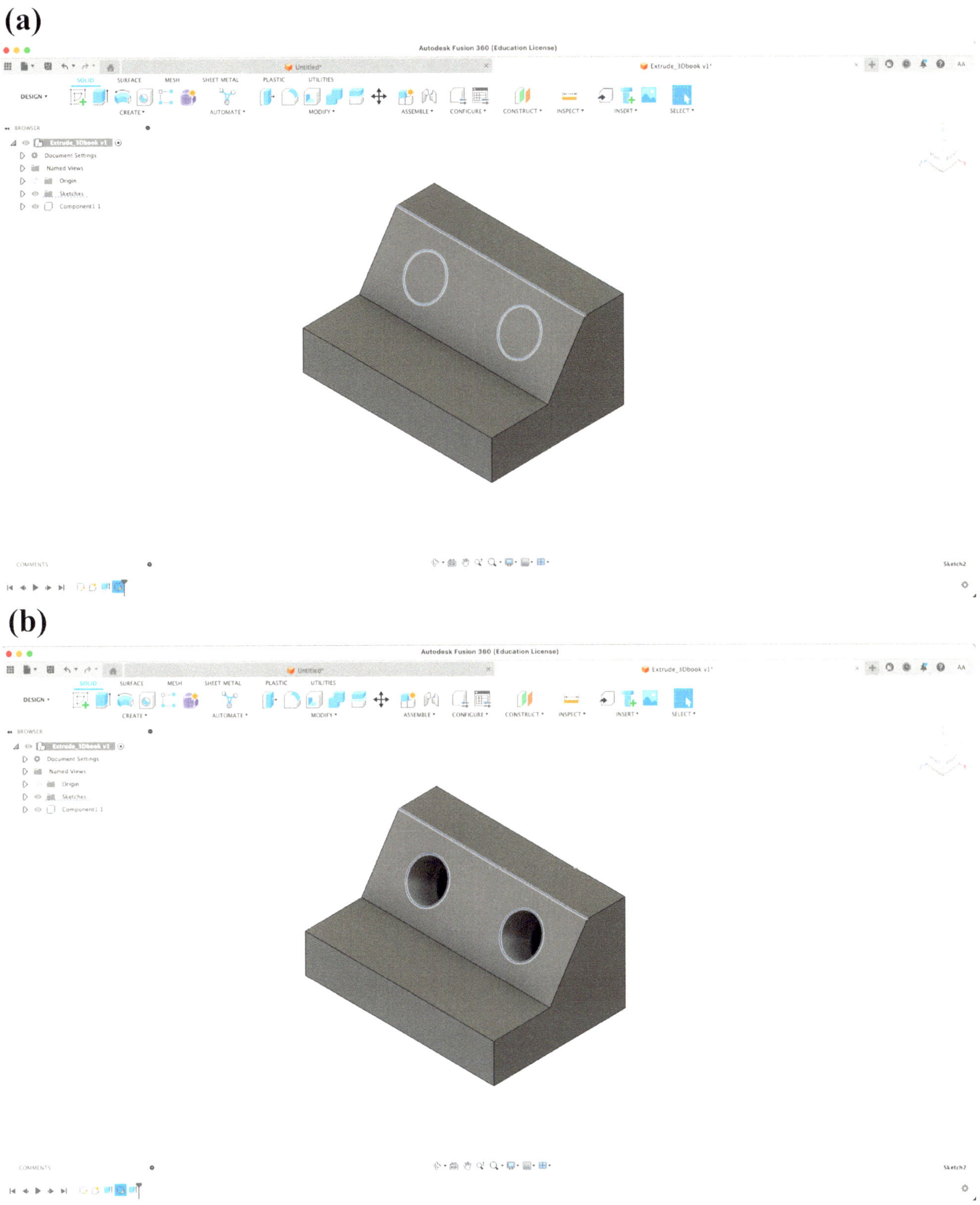

Fig. 7 The sloped surface can now be selected for adding further features. **a** adding the circles and **b** extruding into the shape to create the holes

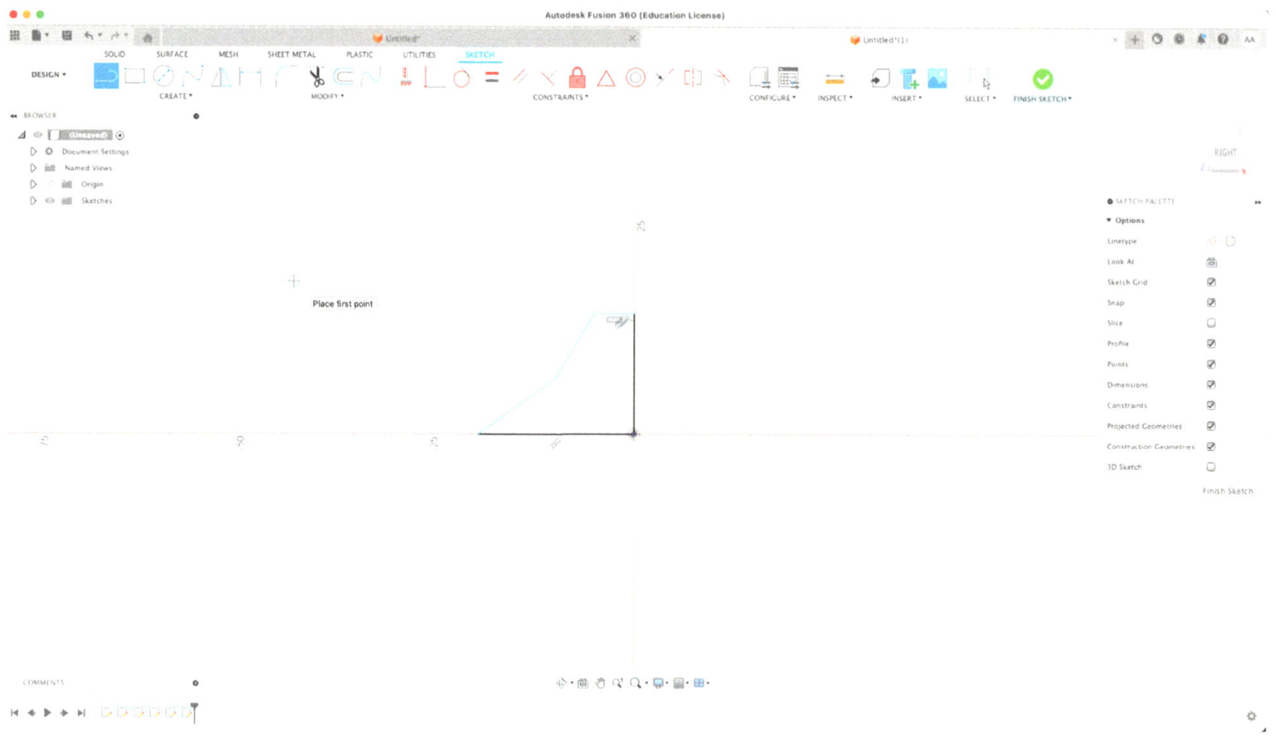

Fig. 8 The sketch and the axis of rotation selection within Fusion360©

7 Sweep

The Sweep command is another useful tool for constructing 3D shapes and features. Consider the solar thermal heater shown in Fig. 12. It would be quite challenging to draw the serpentine pipes and devise their pitch, size and pass number in a straightforward manner had there not been the Sweep command.

In this case, the drawing commences also with a Sketch to be drawn on a plane. The outline of the serpentine pipe is drawn from the thermal and geometric calculations as shown in Fig. 13a.

After this, the pipe diameter, represented as a circle will be added to the starting edge of the pipe network and then *swept* along the path to give it its shape. For this, a plane is constructed using the Construct tool, and this plane is an Off-Set Plane that is parallel to the y–z plane and its location is chosen to be To-Object, which is the starting point of the serpentine pattern as shown in Fig. 13b. A circle having the diameter of the pipe is then drawn on the offset plane as shown in Fig. 13c. Finally, the Sweep command is invoked, where the profile is selected as the circle, and the path is selected as the serpentine pattern of the pipes, resulting in the final shape, shown in Fig. 13d. It is worth mentioning that the serpentine pattern was drawn using the Rectangular Pattern command to equally distribute the straight sections of the pipes, while the elbows are made of 180º arcs.

8 Model Saving, Exporting as. stl File

The next step after a part is constructed is to ensure that its features are amenable to 3D printing. Apart from planning the object features, such as overhangs, curvatures and complex cavities, there are orientations that provide faster tessellation later in the postprocessing steps of slicing the object and creating the .stl file that is readable by the 3D printing apparatus. A facile solution such as the one that was shown in Fig. 3d may not always be present, but it is worth the designer's time to inspect the object and plan for subsequent stages of production as early on as possible.

In most CAD software, and Fusion360® is no exception, there are options to export the finalized object as a. stl file,

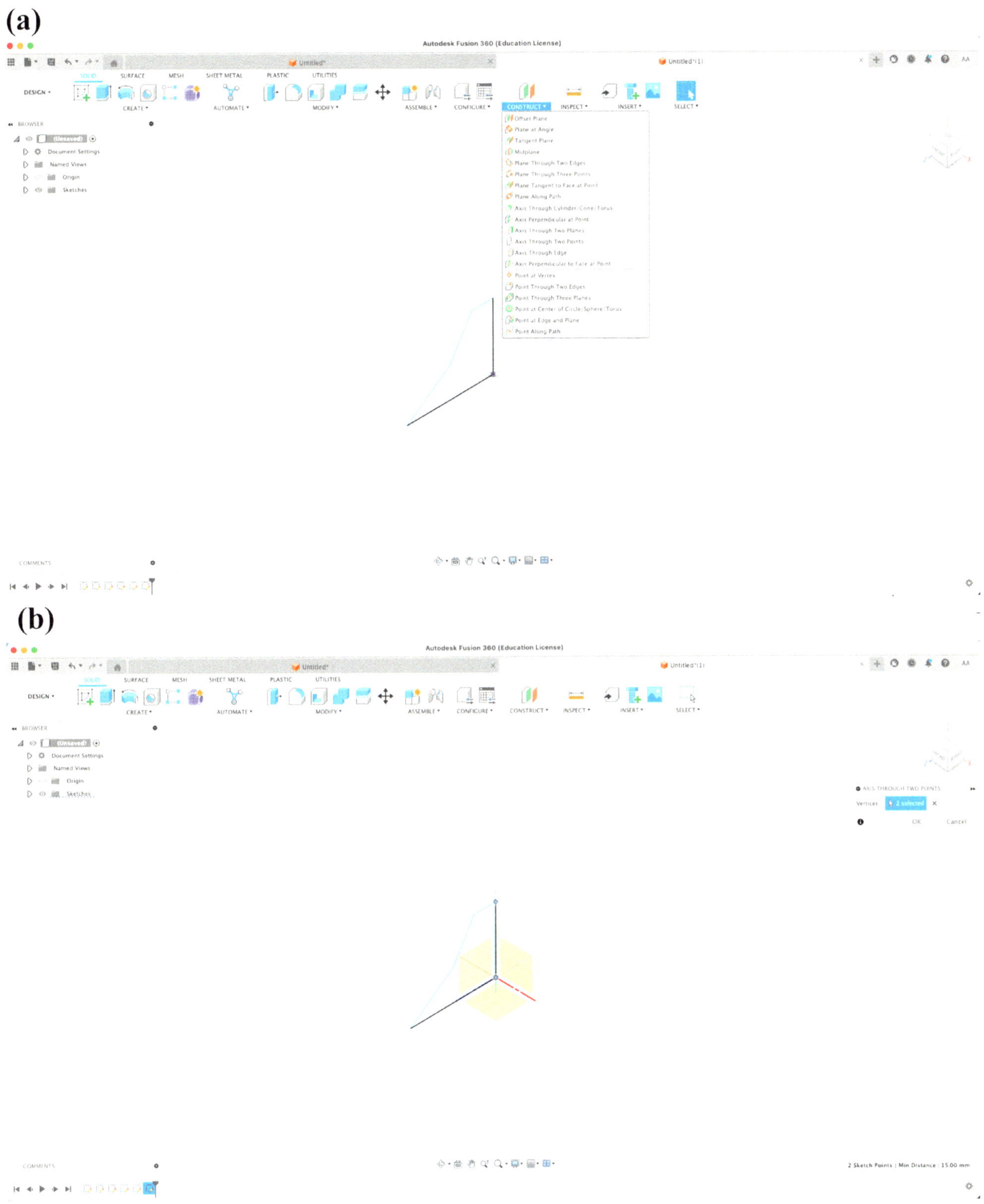

Fig. 9 Revolve operation **a** profile created from the sketch and **b** rotation axis construction

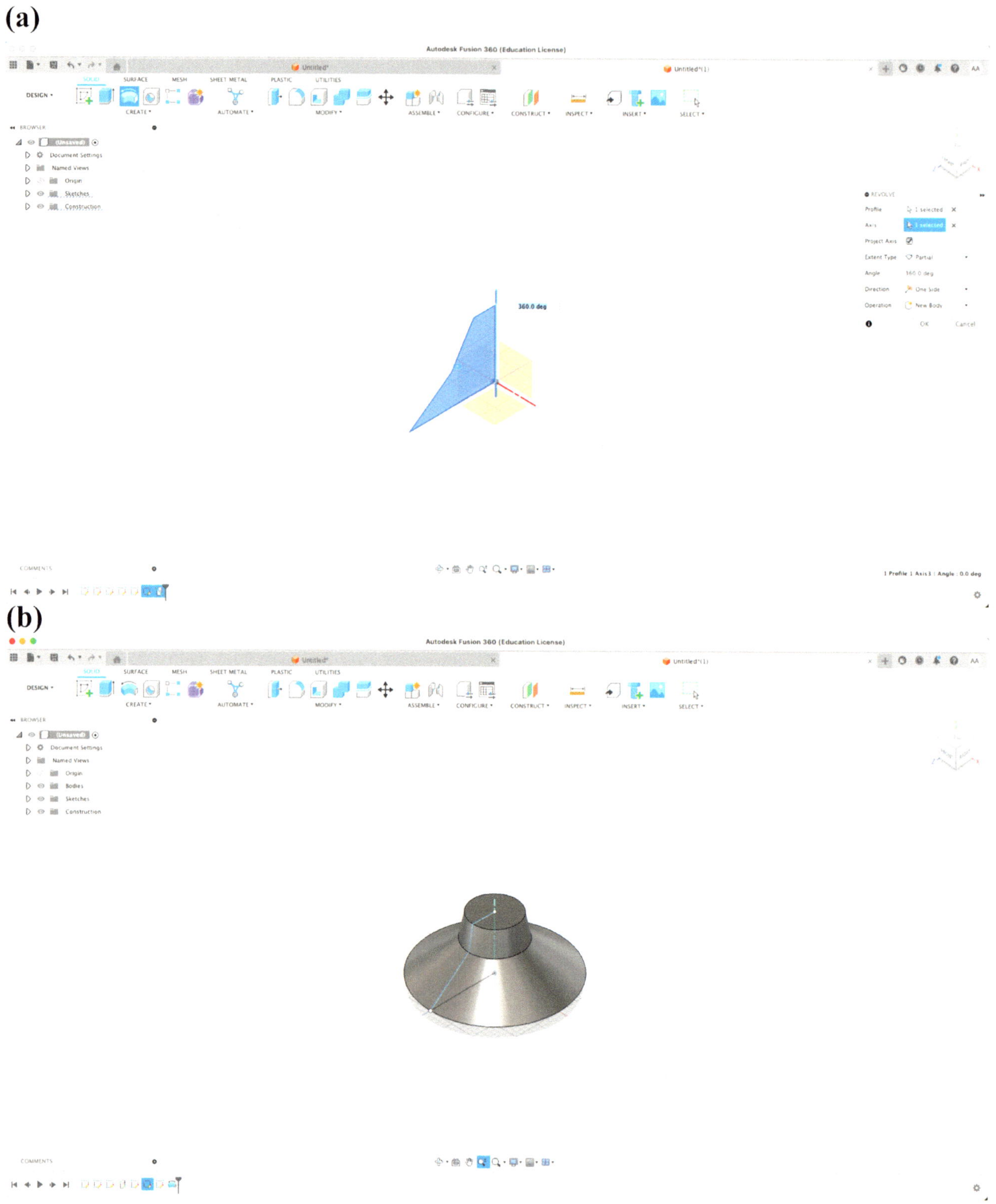

Fig. 10 **a** The revolved command prompts to select the profile and the rotation axis to arrive at the **b** final revolved shape

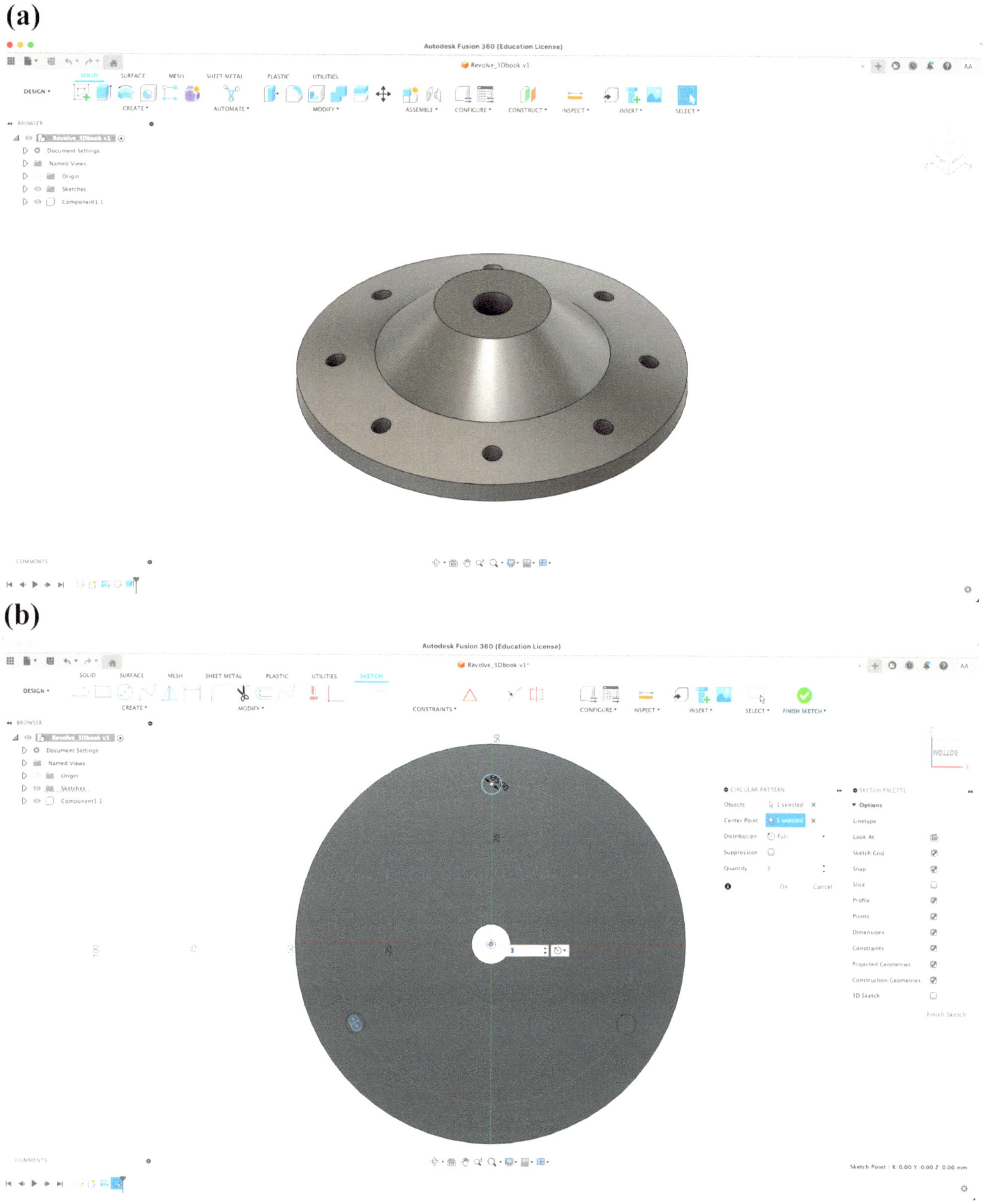

Fig. 11 Adding features to the revolved shape is also easy, where **a** a pattern of holes is added to the surface by drawing a circle and **b** using the circular pattern command to add as many circles as needed then extruding them to create the holes

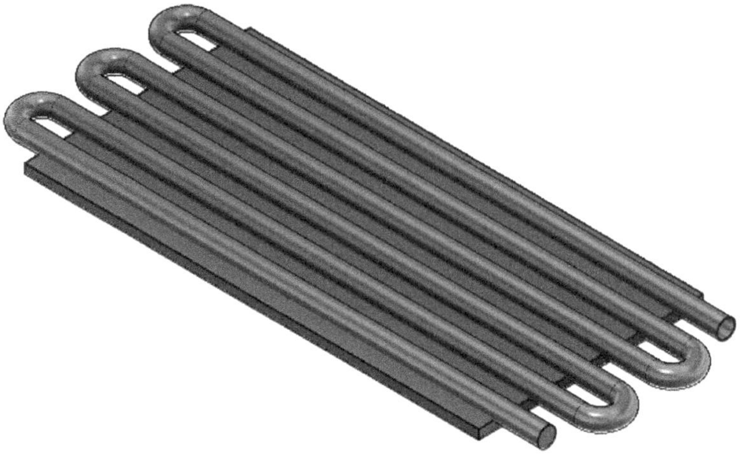

Fig. 12 Solar thermal collector with serpentine pipe arrangement

but the user should be aware of the laminations of the 3D printer, such as nozzle diameter, extrusion temperature, curing parameters, dimensional stability and most importantly, available print volume. If keeping dimensional accuracy is crucial for the user, then a bigger object can be divided into two or more halves and then assembled after the printing process is complete. Otherwise, the Scale option in the CAD or 3D printing software can be used to limit the size of the object. It is interesting to see the way the files look after they are converted into. stl format, and the tessellations are shown in Fig. 14a–c. The number of layers can be modified with more layers and tessellates. This usually happens at the expense of the export time and file size. Thus unless there is a specific requirement and goal, it is best to keep default settings for. stl files.

9 Preparation for 3D Printing

The object file now is ready to be printed, and usually, a printer-specific plug-in or software is required to translate the generic geometrical information into machine code. In the authors case, the printer used is a FlashForge® Creator 3 Pro (Check the company website for the specifications https://www.flashforge.com/product-detail/flashforge-creator-3-pro-3d-printer) and the software used is FlashPrint®. The software is versatile, intelligent and powerful enough to handle professional jobs as shown in Fig. 15a. It includes options to manipulate the geometry (scaling, cutting and rotating) as well as to add supports both manually and automatically. The supports include straight supports and branched ones where the user can control their size and geometric shape as shown in Fig. 15b. The supports are shown in Fig. 15c.

It is also worth mentioning that if the orientation of the part is changed, as shown in Fig. 15d, the need for support is eliminated altogether. The shape was rotated 90° clockwise around the x-axis, with the walls of the openings being in the build direction and hence not requiring vertical support.

10 3D Printing and Evaluation

The printed parts turned out as shown in Fig. 16. The surface quality and fineness of the end product depend on several factors. These factors can be divided into (i) materials, (ii) process and (iii) equipment-related factors.

The material aspect refers to the type of filament or resin used for the print. The materials have a set of parameters where the deposition/curing can take place optimally. As polymeric materials, filaments and resins are sensitive to their environment and storage conditions, and often have a finite shelf life, after which the materials should be discarded and replaced with new materials.

The process aspects refer to process parameters such as nozzle temperature, deposition/scanning speed as well as presets from the printing software such as the filling ratio of the build, whether a raft is included and the type and density

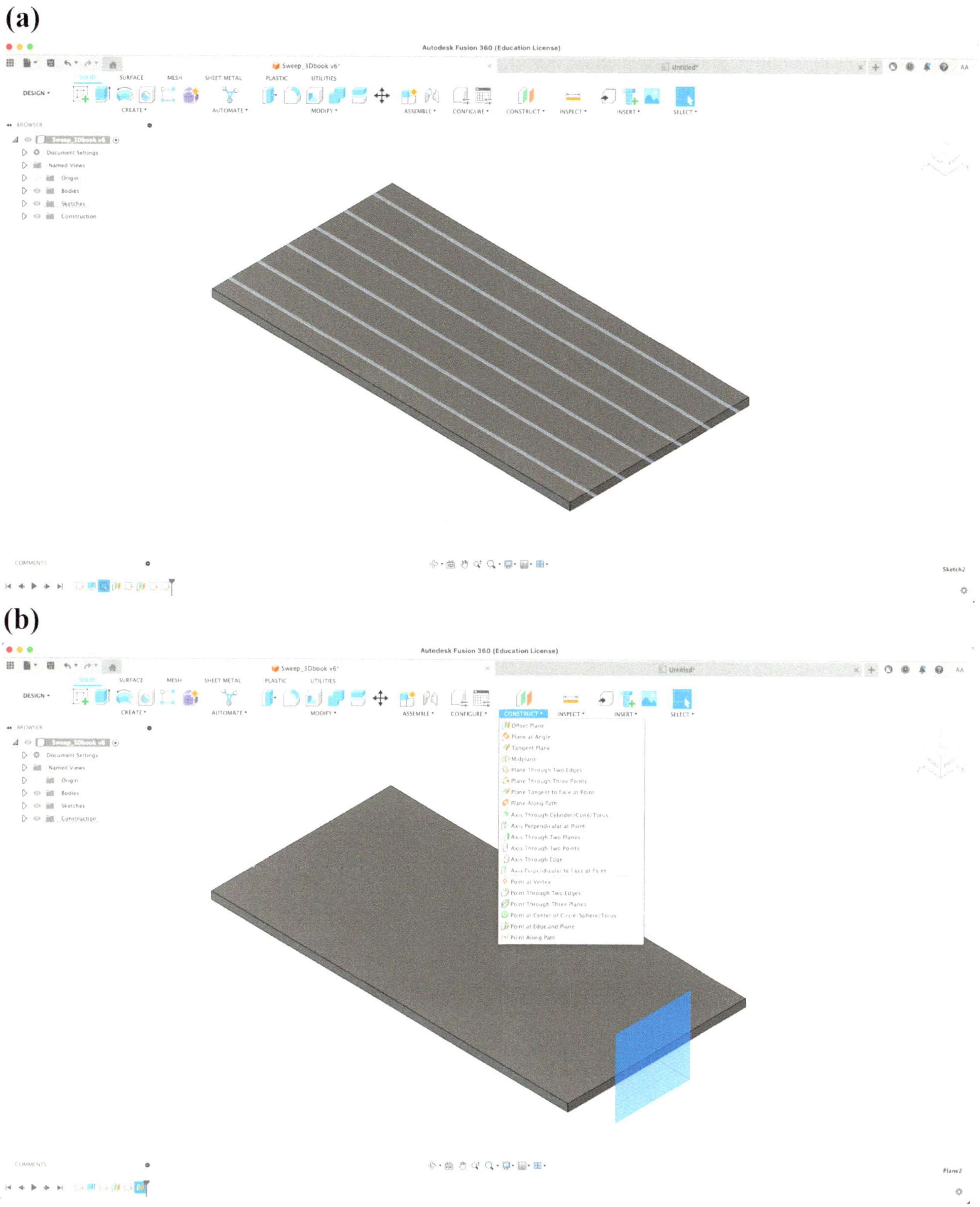

Fig. 13 The process of drawing the solar thermal collector using the Sweep function, **a** first a path is drawn using the Sketch command, then **b** a plane is constructed to draw the perpendicular circular profile that matches the pipe's external diameter in **c**, and **d** how the final drawing looks after Sweep command is executed

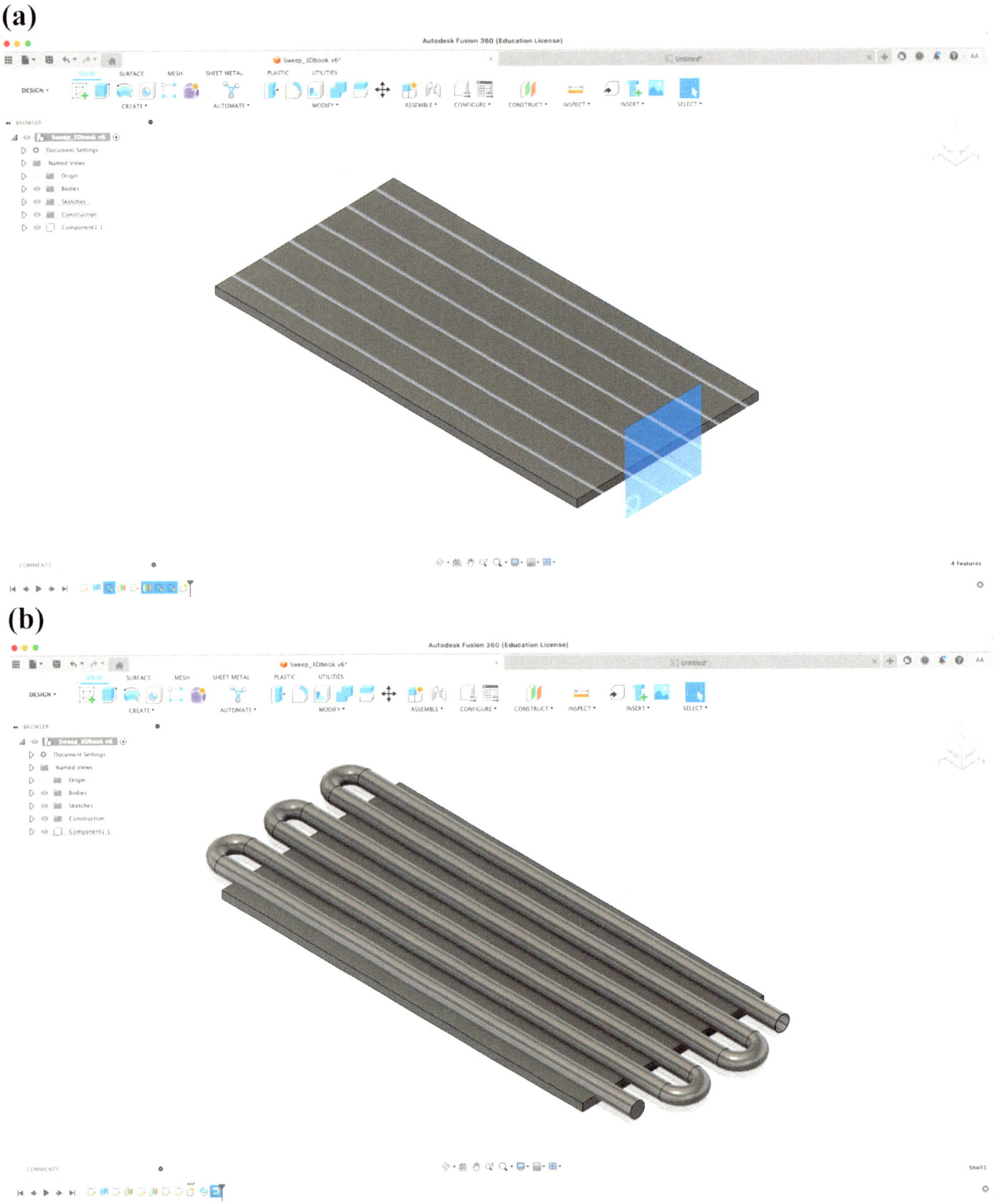

Fig. 13 (continued)

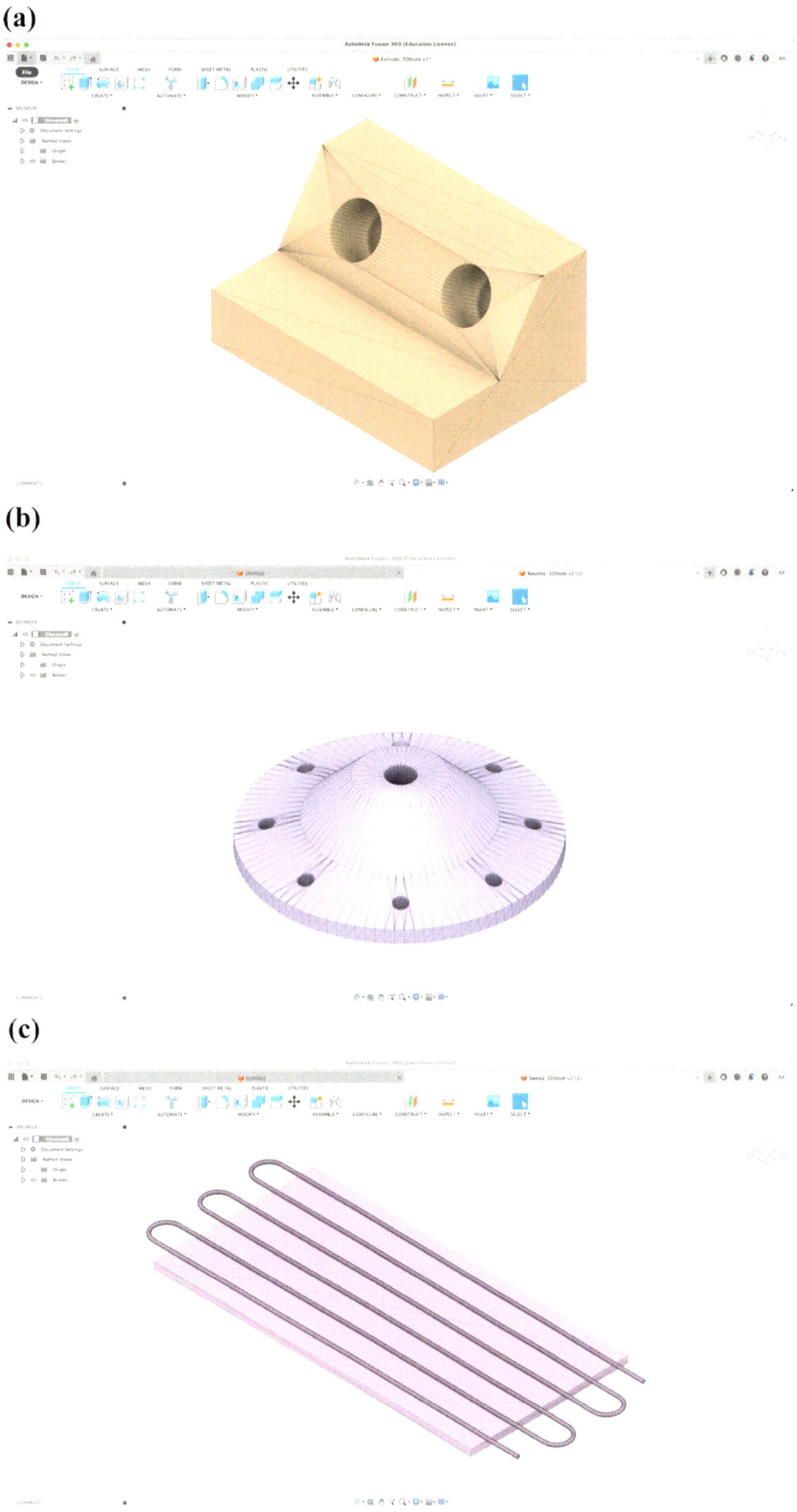

Fig. 14 The .stl files generated from different geometrical construction tools **a** Extrude, **b** Revolve and **c** Sweep

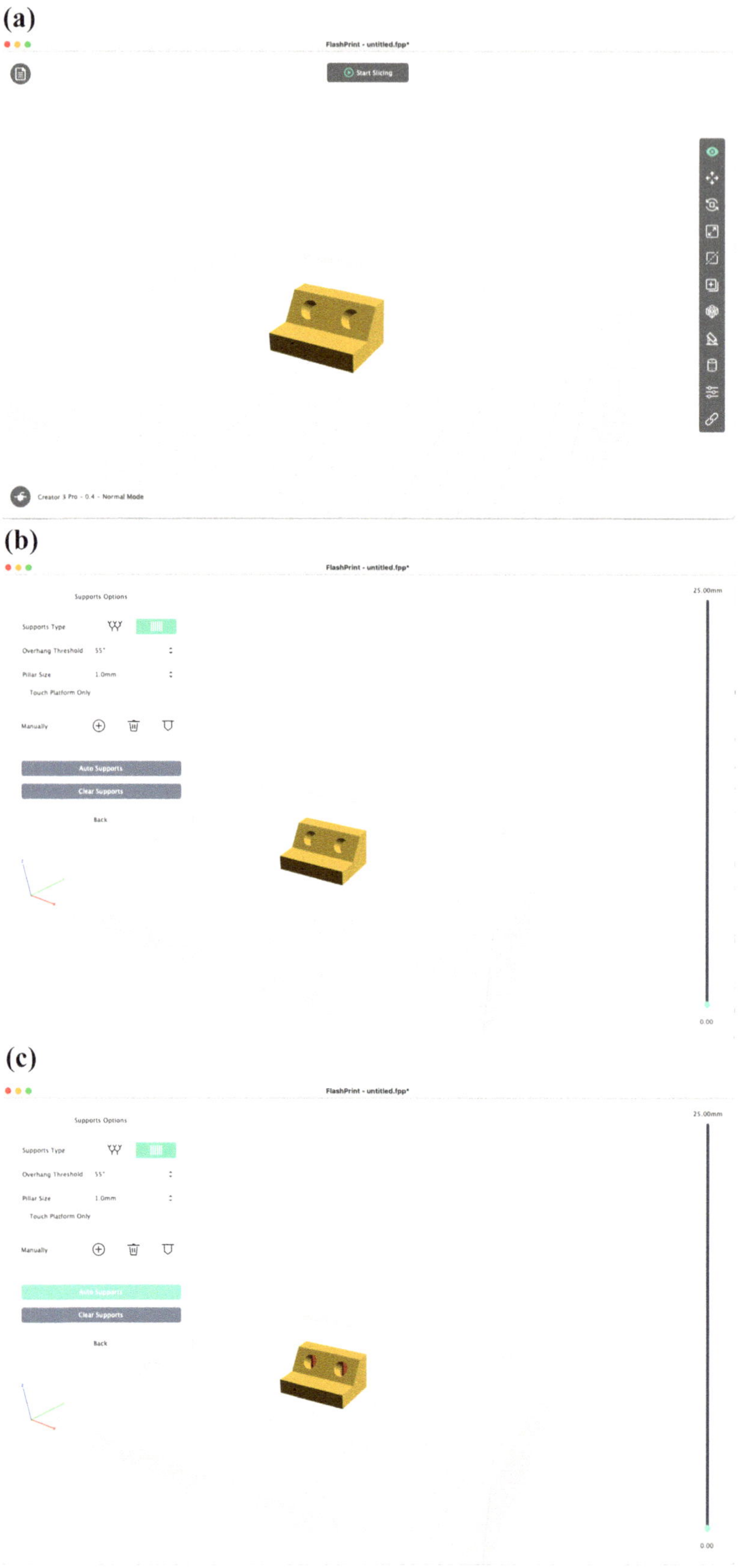

Fig. 15 The process of converting the idea into a product: **a** the .stl file is imported into the software driver of the 3D printer, **b** adding required building supports, **c** supports (in red inside the holes) are added, **d** changing the orientation may result in a build sequence that does not require supports, which is preferable and **e** the tesselated/sliced preview of the part with layers, structure breakdown and time estimation

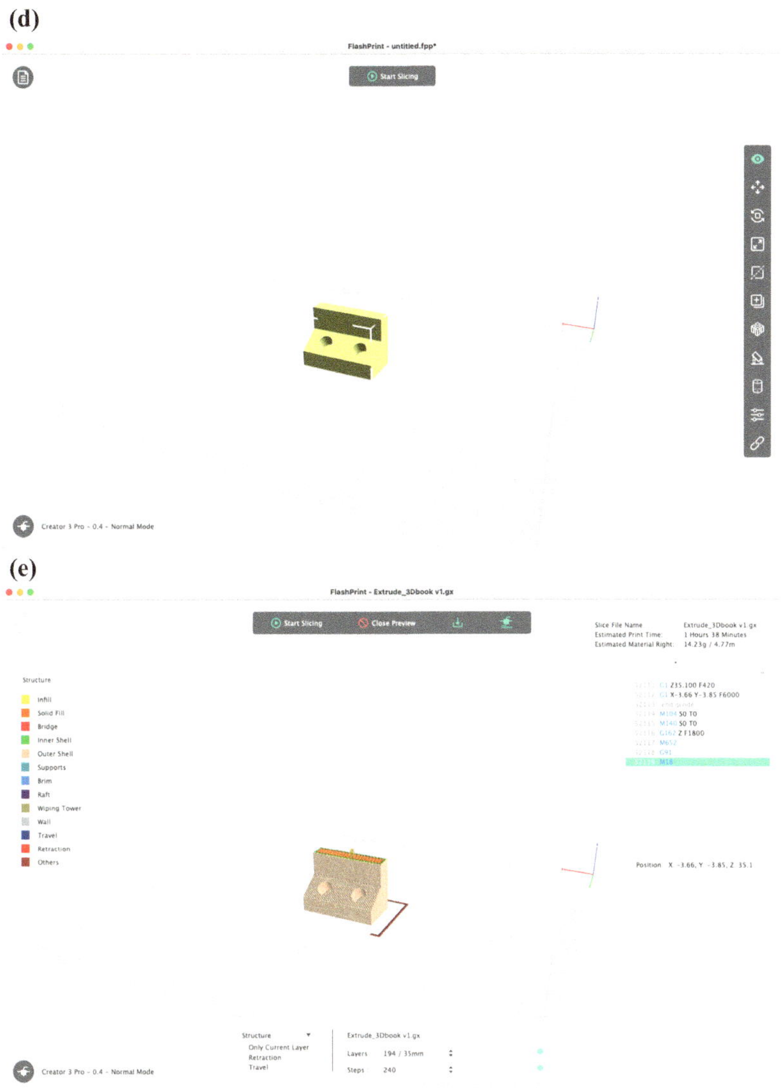

Fig. 15 (continued)

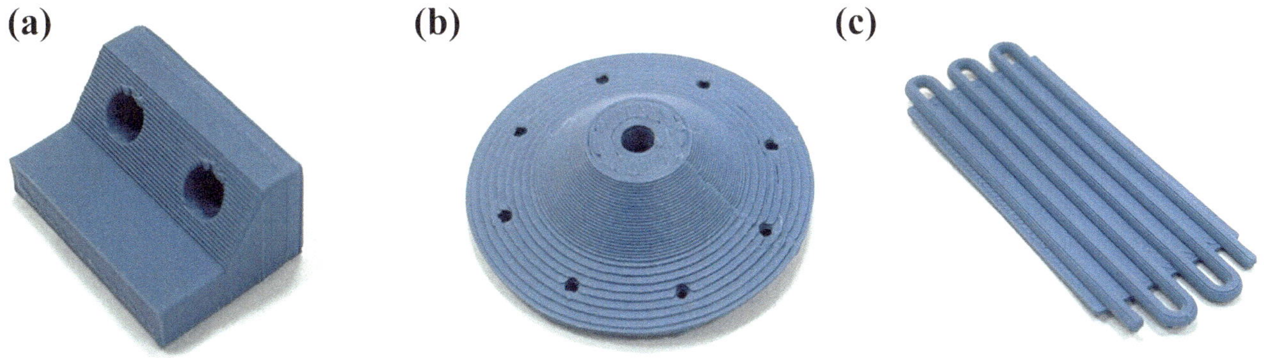

Fig. 16. 3D printed parts for the **a** Extrude, **b** Revolve and **c** Sweep commands

of the supports. More supports and denser support additions are detrimental to the cleaning process once the printing is done.

Finally, the equipment choice is important in terms of resolution, speed, size and filament dispensing. The market is filled with many different options that must match the expected print quality and quantity.

Reference

How People Worked Before AutoCAD. (n.d.). Retrieved July 26, 2024, from https://www.wmeglobal.com/how-people-worked-before-autocad/

Additive Manufacturing Technologies

Abdul Hai Alami, Ayman Mdallal, Mohamad Ayoub, Adnan Alashkar, and Ahmad Yasin

Abstract

The focus of this chapter is on different processes used in manufacturing tangible products. Moreover, the chapter showcases (i) machining (material removal) and (ii) additive manufacturing techniques. It sheds light on key characteristics involving machining techniques such as turning, drilling, and milling, as well as compatible materials. Then, conventional additive manufacturing techniques, which mainly utilize polymer-based filaments, resin, and metal powder, are described, highlighting their advantages over long-used machining techniques, in a rapid-prototype-demanding environment. Finally, commercial 3D printers and materials are presented to the reader, offering guidance in choosing appropriate tools for their applications.

A. H. Alami (✉) · M. Ayoub · A. Yasin
Sustainable and Renewable Energy Engineering Department, University of Sharjah, 27272 Sharjah, United Arab Emirates
e-mail: aalalami@sharjah.ac.ae

M. Ayoub
e-mail: mohamad.ayoub@sharjah.ac.ae

A. Yasin
e-mail: ahmad.yasin@sharjah.ac.ae

A. Mdallal
Sustainable Energy and Power Systems Research Centre, RISE, University of Sharjah, 27272 Sharjah, United Arab Emirates
e-mail: Ayman.Mdallal@sharjah.ac.ae

A. Alashkar
PhD Program in Materials Science and Engineering, American University of Sharjah, 26666 Sharjah, United Arab Emirates
e-mail: b00028197@aus.edu

1 Available Manufacturing Technologies and Materials

Manufacturing technologies, a continually evolving sector, use many methods to convert raw materials precisely and efficiently into final products. A list of these methods is shown in Fig. 1. For decades, casting, forming, and machining have been essential to the manufacturing sector. Cutting-edge technology like Additive Manufacturing (AM) has changed the sector. These manufacturing methods and production processes play an important role in the automotive, aerospace, electronics, and healthcare industries (Fidan et al., 2023). In 2021, the manufacturing sector in the United States alone employed over 11.2 million workers (UNIDO Statistics Data Portal, n.d.). Additionally, the global automobile industry has an average yearly revenue of over 2750 billion euros, accounting for 3.65% of the world's gross domestic product (Saberi, 2018). Furthermore, based on data from the World Bank, the manufacturing sector had an annual growth rate of 8.1% in 2021, and a value of 16.55 trillion US dollars in 2022 (Manufacturing, Value Added (Current US$) | Data, n.d.).

Once a material is selected to be used in making a product, a manufacturing process must be devised to produce the part. All available manufacturing processes have energy ratings that will match the change to be accomplished to arrive at the final part. This energy can be (i) thermal, utilized to melt the material; (ii) compressive forces applied using mechanical, hydraulic, or pneumatic actuators that are coupled with respective motors that have the required horsepower; or (iii) applied via the tip of a cutting tool that applies sufficient force over a point area to generate high stresses (stress is the force over area) that allow instantaneous material degradation and thus removal.

One of the most representative and basic relationships in materials engineering is generated by performing uniaxial tensile or compression tests. A specimen is placed between

A. H. Alami (ed.), *Sustainable Additive Technologies and Innovative Manufacturing Techniques*, Advances in Science, Technology & Innovation, https://doi.org/10.1007/978-3-031-73657-5_4

Fig. 1 Manufacturing processes and mechanical properties

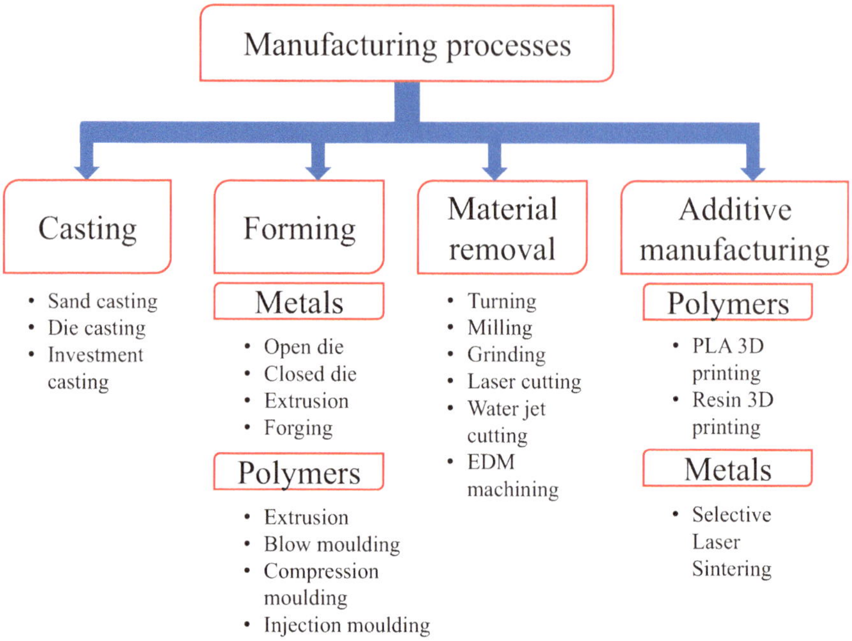

the jaws of the testing machine as these jaws move either away or towards each other, producing tensile or compressive stress within the specimen, as shown schematically in Fig. 2. The load is applied and measured through a load cell that is coupled with the machine, and the rate of motion of the jaws is plotted against the load. The machine jaws usually travel at a predefined constant rate, and if the load cell readings were taken against time, the time stamp can be multiplied by the travel rate to get the deflection value.

The plot is later standardized by removing geometrical information pertinent to the size of the sample. This enables the acquirement of generalized mechanical properties that are valid for the tested material, regardless of the dimensions. This is accomplished by calculating the stress which is represented by the quotient of the load and cross-sectional area of the specimen and dividing the deflection by the original gauge length of the specimen. A generic typical curve is shown in the inset of Fig. 2.

The importance of the stress–strain curve lies in two main points. The first is that it indicates stress values mile stones that are important for identifying the mechanical properties of the material, such as the yield point, where any stresses beyond this point would result in permanent (plastic) deformation of the material. Below the yield point, if the load is removed from the part, the part is expected to return to its original shape and dimensions, and the deformation is dubbed (elastic) deformation. As the loading continues, as shown in Fig. 3, the material reaches the ultimate tensile strength point, which is usually taken as the strength of the material. This is the highest stress level that the material can handle on its way to fracture. It is usually accompanied by visible necking during tensile tests on the specimen as internal faults and voids coalesce within the material, and the load that the material can handle becomes progressively less until the fracture point, which usually occurs at the maximum displacement (strain) value.

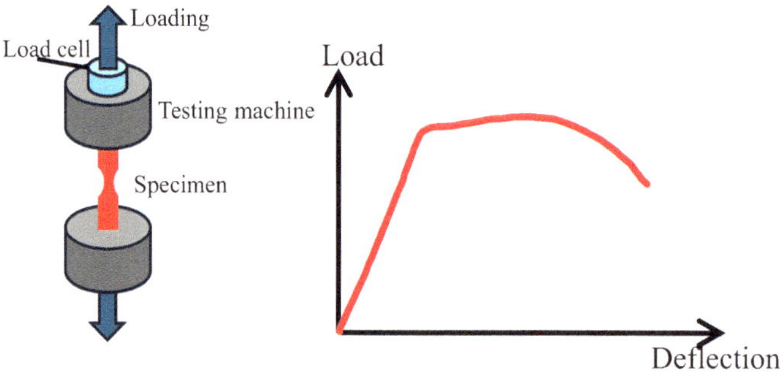

Fig. 2 Depiction of the uniaxial testing procedure

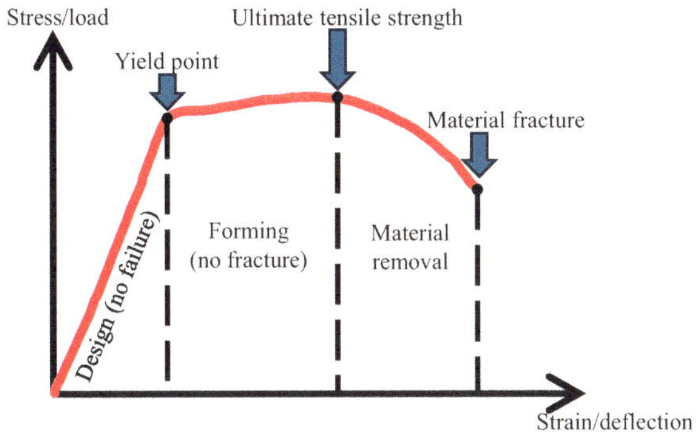

Fig. 3 Typical and generic stress–strain diagram

The stress–strain diagram also provides important information about the energy required to manufacture the part. The area under the stress–strain diagram represents the energy required per unit volume of material and is therefore independent of the tested specimen size. This can be verified using a simple unit check:

$$\text{Area under the ; curve} = \text{stress} \times \text{strain} = \left(\frac{N}{m^2}\right) \times \left(\frac{m}{m}\right) = \frac{N \cdot m}{m^3} = \frac{\text{Joule}}{\text{Volume}} \tag{1}$$

This argument means that by obtaining the stress–strain diagram, a designer can select the range of loads that should be applied to the material to arrive at the final required shape. This implies that for a manufacturing process to succeed in changing the shape of the material, the process should apply loads that will result in stresses in the material that will take it beyond the yield point. Otherwise, the material will return to its original shape and dimensions once a sub-yielding load is removed. On the other hand, such loads should not cause the material to fracture or disintegrate while being formed, which entails staying below the ultimate tensile stress. This translates into several provisions in a forming process that uses dies mounted on compressive presses to force a material to conform to the cavity shape in the die. These provisions include heating up the material so that it would require lower loads to form without fracturing, using successive dies that shape the material in sequential steps using a family of dies, or using heat treatment between forming steps to heal the material back into a more elastic form.

Finally, sometimes material removal is the goal of the manufacturing process. From the stress–strain diagram, the range of such operations starts at the necking that ushers in the commencement of disintegration of the material at the ultimate tensile stress. This also indicates that the stress value must be as high as possible, and this is attained either by increasing the loads (not always possible) or by reducing the cross-sectional area. Most material removal processes use hard and sharp tools mounted on strong motors. The tool's cutting point is the area where the material experiences a stress level that easily reaches the fracture stress value in the stress–strain diagram, enabling the material to be chipped away. The tool should be harder than the material and has a sharpened tip for optimum results. It is evident that the stress–strain diagram, although extremely facile to obtain, is a goldmine of information on the most suitable manufacturing process for a certain material, as well as the required levels of loads that would generate the forming stresses within the material. In this chapter, there will be a focus on additive manufacturing as well as material removal processing since the latter is the antithesis of the former. This being said both share many aspects of preprocessing (CAD files, tessellation, etc.) as well as process control technologies and even post-processing steps. For example, material removal and additive manufacturing technologies use computer numerical control (CNC) for path planning and material handling, which is mandatory for both applications. CNC requires detailed computer-based files to plan the machine path and execute the process. In case of material removal, a g-code file has to be generated that guides the cutting tool over the original part, while the.stl files are meant for a depositing nozzle of a 3D printer as shown in Fig. 4.

In the early 2000s, CNC transformed the manufacturing and machining industries with high production rates, tolerances, and accuracies. CNC machining utilizes advanced computerized technologies to precisely control the motion and operations of machine tools, with precision reaching $\pm$ 0.005 mm (Precision CNC Metal Machining—Metal Cutting Corporation, n.d.) instead of relying on their conventional, manually controlled counterparts. Intricate parts, prototypes, and refined products are now produced with an unmatched level of precision and repetition. Since the processes start with a CAD file, as shown in Fig. 4, this allows collaboration (that can take place online in different countries and time zones), archiving the part for quality control and further development, as well as running a path simulation or part slicing to predict the time a part needs to be completed and the routing of the tool/nozzle that can be optimized either for time or for accuracy (Eladawi et al., 2003; Li et al., 2015). Monitoring systems and sensors are needed in these machines to guarantee adherence to operational standards and minimize errors during the process of removing materials, but they add to the price tag. Certain processes need additional post-processing steps, like polishing or heat treatment, to achieve the desired dimensional accuracy and surface finish.

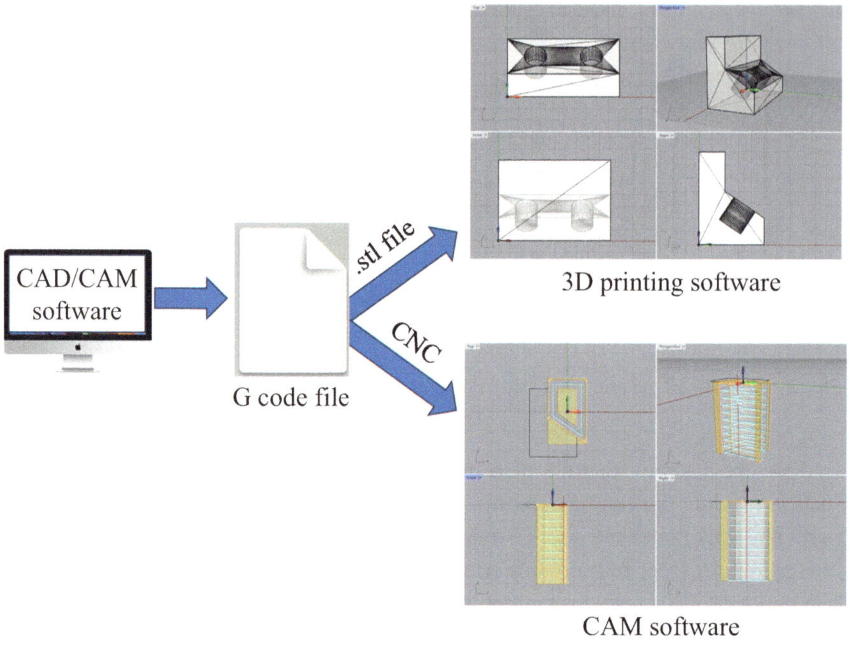

Fig. 4 Two options for a CAD file after the CAM conversion into a g-code that drives either a 3D printing nozzle or a CNC cutting tool

2 Machining (Material Removal Processes)

This section covers conventional and CNC-based material removal processes. The purpose is to highlight the contrast of these processes to additive manufacturing technologies.

2.1 Turning Process

In this process, a workpiece is rotated on a spindle while a cutting tool is used to remove material to obtain specific shapes and dimensions. It is frequently employed for cylindrical components, such as shafts, rods, and disks (Kumar et al., 2019). The selection of materials for turning varies depending on the application, covering metals such as steel, aluminum, and brass, as well as polymers and wood. Turning is a cost-efficient technique commonly used to manufacture simple circular components.

2.2 Milling Process

Milling is a highly adaptable machining technique employed to eliminate material from a workpiece by moving a revolving cutter along its surface (Kumar et al., 2019). Through this method, numerous shapes, including flat surfaces, slots, and complex three-dimensional structures can be fabricated. Milling machines can handle a variety of materials, ranging from aluminum and steel to polymers and composites.

2.3 Drilling Process

Drilling is a machining method where holes are formed in a workpiece by means of a rotating cutting tool known as a drill bit (Kumar et al., 2019). This method is mainly utilized for metals, woods, and plastics allowing for its utilization in construction, aerospace, and manufacturing.

2.4 Sawing

Sawing is the process of removing material by using a thin teethed-metal sheet (forming a saw) that moves back and forth. The teeth of the tool are typically composed of carbide or high-speed steel, and every one of them removes a small amount of material with each stroke (Kumar et al., 2019).

2.5 Wire Electrical Discharge Machining (Wire EDM)

An innovative method of manufacturing that uses electrical discharges to shape and cut materials. The process utilizes a thin wire acting as an electrode to produce detailed and accurate cuts in the workpiece without contacting the material (Patel et al., 2014). Wire EDM is commonly used for components that demand high precision and complex forms and utilizes conductive materials such as metals and alloys.

This technique employs a thin wire, typically made of copper, brass, or zinc, with a diameter ranging from 0.05 to

0.3 mm. This wire is utilized to cut through materials that are electrically conductive, using a process called thermal erosion. Throughout the process, a series of electric sparks are produced between the wire and the piece, causing the work and tool material to melt and vaporize. During the cutting process, a constant stream of dielectric fluid is circulated around the wire to efficiently remove molten material from the spark zone as debris. An oscillating voltage is employed to produce a current between the electrodes. Once the voltage supplied hits a critical threshold, a plasma channel is formed, leading to the initiation of a spark. Therefore, there is a significant increase in temperature at a specific location, causing the substrate material to liquefy and transform into vapor. The plasma channel stops when the pulsating power supply is deactivated, resulting in a sharp drop in temperature. The dielectric facilitates the removal of particles from the machined surface (Ho & Newman, 2003). A wire is continuously supplied without making any touch with the workpiece. An automatic maintenance of a clearance ranging from 0.025 to 0.05 mm is ensured between the workpiece and the wire, in accordance with the machining parameters (Rakshit & Das, 2019).

2.6 Laser Beam Machining

Laser machining is a very accurate manufacturing technique that uses focused laser beams to remove material from a workpiece. The process involves directing a laser beam with high intensity onto the surface of the material, resulting in localized heating and vaporization, melting, or ablation of the material. The controlled thermal energy allows for the accurate removal of material while minimizing damage to the surrounding areas (Dubey & Yadava, 2008).

2.7 Abrasive Water Jet Machining

The adaptability of abrasive water jet (AWJ) technology has led to its widespread use in manufacturing industries for machining many materials, including both metals and non-metals. The reason for preferring AWJ machining is its ability to produce less heat in the cutting area, which is especially beneficial when working with hard materials. Moreover, AWJ machining demonstrates the ability to efficiently cutting, drilling, and milling a variety of materials, ranging from metals to non-metals, composites, and ceramics (Folkes, 2009). AWJ exhibits a greater pace of removing material in comparison to traditional methods such as Wire EDM, while also providing enhanced surface integrity as compared to laser machining techniques. This allows for the precise cutting of thick components, typically up to around 250 mm, depending on the exact material properties. Significantly, it avoids heat deformation in work materials while enabling the precise shape of components. This technique is additionally characterized by its use of low cutting forces on work materials, resulting in improved dimensional accuracy and reduced deformation. In addition to conventional machining applications, the applicability of AWJ technology can be extended to jobs such as rock drilling and surface cleaning (Natarajan et al., 2020).

3 Additive Manufacturing

In this section, the focus will shift to the most common additive manufacturing technologies. These are the technologies at the heart of 3D printers that are available off-the-shelf and in the market.

3.1 Fused Deposition Modeling (FDM)

FDM is a commonly utilized method for prototyping and producing functional components utilizing thermoplastic materials (Cicala et al., 2017). The process, shown in Fig. 5, involves melting a strip of plastic and depositing it in layers that follow to construct the final 3D item. FDM is known for its flexibility, cost-efficiency, and accessibility (Daminabo et al., 2020). The materials utilized in FDM include a range of thermoplastics, such as ABS, PLA, and PETG.

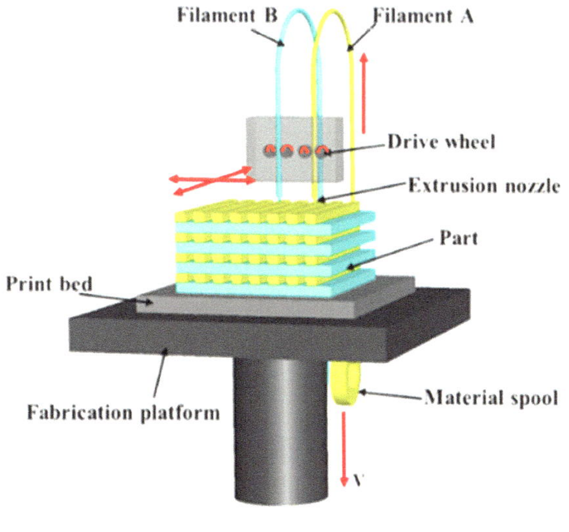

Fig. 5 Fused deposition modeling—adapted from (Wang et al., 2017) (with permission no: 5814801374925)

3.2 Stereolithography (SLA)

A method that uses a liquid resin solidified by UV light or a laser beam to produce 3D objects one layer at a time. The process is displayed in Fig. 6. SLA is recognized for its ability to produce items with polished surface textures, thus it is generally employed in the aerospace, automotive, and healthcare industries. The materials utilized in SLA include a range of photopolymer resins that possess unique features such as transparency, flexibility, or high-temperature endurance (Mouzakis & Mouzakis, 2018; Vladić et al., 2022).

3.3 Selective Laser Sintering (SLS)

SLS uses a high-intensity laser to selectively fuse powdered materials, typically polymers or metals. This technology is valued for its versatility, allowing for the creation of functional prototypes and end-use components with intricate shapes and outstanding strength. Materials used in SLS include various powders, such as nylon and polyamide, as well as metals like aluminum and stainless steel (Mouzakis & Mouzakis, 2018).

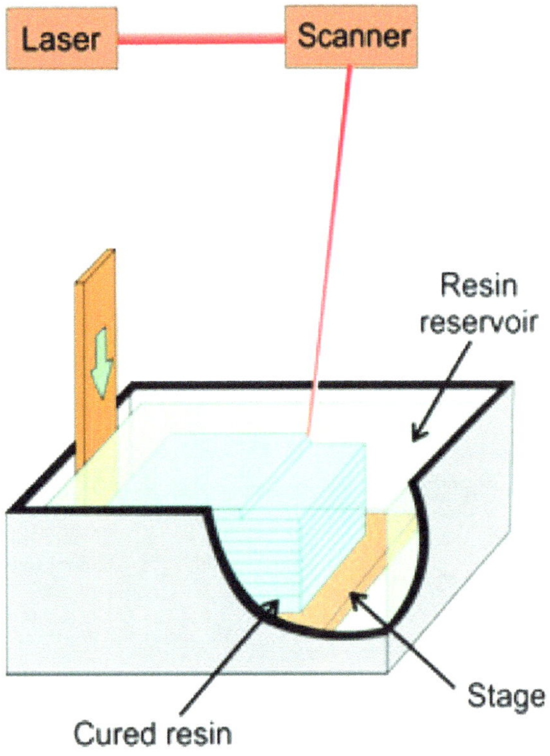

Fig. 6 Stereolithography (SLA)—adapted from (Moritz & Maleksaeedi, 2018) (with permission no: 5814810228452)

3.4 Electron Beam Melting (EBM)

EBM utilizes an electron beam to precisely heat and fuse metal powder one layer at a time, resulting in the production of 3D objects. EBM is frequently employed in the aerospace and medical sectors. EBM is mainly utilized to construct parts from metal powders such as titanium and cobalt-chrome alloys. Starting with a powder bed, CAD data is used to build the 3D target geometry. The AM technique may fail due to powder electrostatic charge. However, EBM offers unique material fusing and melting solutions due to its high velocity limit (up to 105 m/s), inertia-free movement and focus features, and quick electron beam movement capacity (Körner, 2016; Uçak et al., 2022).

3.5 Binder Jetting (BJ)

BJ involves the utilization of a liquid adhesive to selectively bind powdered materials layer by layer. This method enables the production of multifaceted and intricate objects using a variety of materials. Binder Jetting is employed for the purposes of prototyping and small-scale production. It utilizes a variety of powders, including metals, ceramics, and sand. Binder jetting provides a comparatively faster construction pace in comparison to material jetting. The building material is evenly distributed around the platform using a roller. Then, the inkjet printer applies the binder, primarily in its liquid form, onto specific regions of the build material. This process effectively connects the particles of the substrate and creates a new layer. The process is repeated by lowering the build platform and fitting the following layer (Srivastava et al., 2022).

3.6 Metal Powder Bed Fusion

Laser powder bed fusion is a method that employs a high-intensity laser to precisely heat and merge metal powders in a layered manner, resulting in the formation of solid metal components with no porosity (Rezaeifar & Elbestawi, 2022), such as shown in Fig. 7. This technique is widely used in the aerospace, medical, and automotive sectors to create intricate and high-quality parts (Dev Singh et al., 2021). The materials utilized in Metal Powder Bed Fusion include metal powders, ranging from stainless steel to aluminum, and titanium alloys.

3.7 Digital Light Processing (DLP)

DLP employs digital micro-mirror device (DMD) technology. The photosensitive material undergoes localized

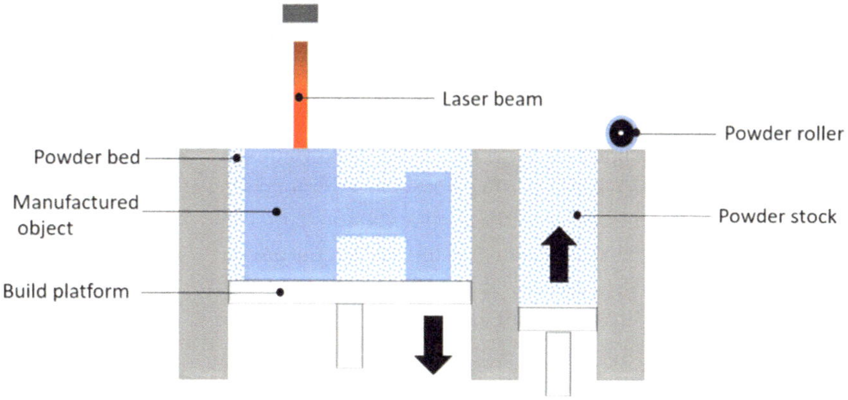

Fig. 7 Powder bed fusion—adapted from (Pal & Basak, 2022) (Open Access)

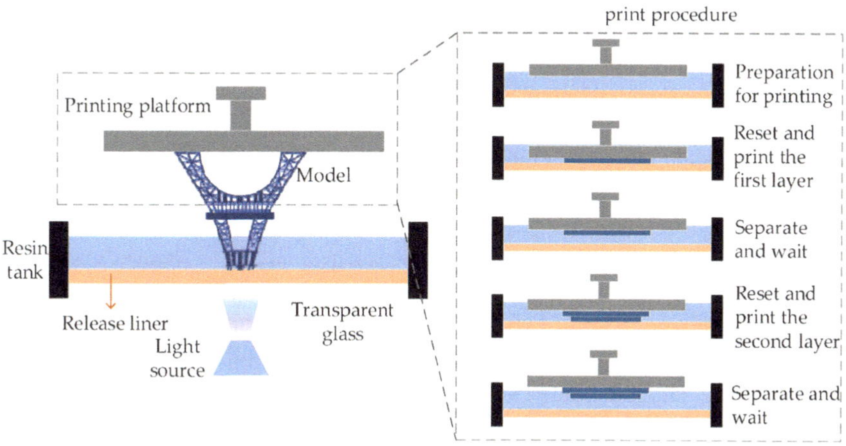

Fig. 8 Digital light processing—adapted from (Jiang et al., 2023) (Open Access)

polymerization and as a result, a stack of layers is formed through the projection of a 2D layer image from a DLP source in a back-to-back manner, as shown in Fig. 8. The images are composed of bright and dark pixels created by micron-sized mirrors on the DMD, which determine the resolution of the polymerized layer in the XY plane. This technology is classified as a vat polymerization process, similar to SLA (Chaudhary et al., 2023). Because it polymerizes the entire layer (slice) of the structure in a single exposure step, it offers reduced build times compared to SLA (Ligon et al., 2017).

3.8 Continuous Liquid Interface Production (CLIP)

CLIP utilizes liquid resin and a continuous liquid interface to construct items layer by layer. CLIP is well-known for its rapid production capabilities and ability to manufacture parts with excellent surface polish. The core concept of the CLIP method is straightforward: UV light projected from below solidifies the photosensitive resin, while the liquid resin in the tank remains fluid due to oxygen inhibition, allowing for continuous curing. A special window at the bottom allows both light and oxygen to pass through. This technology revolutionized production by offering a speed that is 25–100 times faster than traditional DLP 3D printing methods (Quan et al., 2020).

4 Available Printing Technologies

The following section contains a summary of commercially available printer brands along with a breakdown of their concept of operation, specifications, and compatible materials. There are updates and reviews of each of the printers available online with all the pros and cons as well as the specifications.

4.1 Bambu Lab P1S 3D Printer

Recognized for its remarkable speed, precision, and contained working space. With its CoreXY mechanism, this printer can reach speeds of up to 200 mm/s, allowing it to create detailed models like a Benchy in just 18 min. The AMS Hotend is well-known for its exceptional precision and ability to support up to 16 filaments simultaneously. This allows for the production of prints with several colors and materials, showcasing an unmatched level of detail. The design allows for reliable printing using high-temperature filaments such as ABS and PA-Nylon. The airflow in this chamber is controlled by three fans, which maintain consistent material distribution and prevent deformation. Moreover, an activated carbon filter effectively reduces fumes and odors, allowing for its utilization in locations such as homes and educational institutions. The P1S incorporates a built-in LiDAR sensor to automate bed leveling, eliminating the requirement for human calibration, and improving the user experience. In addition, the printer features a removable spring steel PEI sheet that allows for effortless removal of prints. Furthermore, it is equipped with a built-in camera that provides convenient remote monitoring of prints through the user-friendly Bambu Studio application. The key features of this printer are its generous build space of 256 × 256 × 256 mm, a heated bed that can reach temperatures up to 100 °C, a nozzle temperature capacity of up to 300 °C, and its ability to accelerate up to 20,000 mm/s^2 (Bambu Lab P1S 3D Printer P1S / In Stock | Bambu Lab US, n.d.).

4.2 Original Prusa MK4 3D Printer

The Prusa MK4 3D printer shows an integration of high performance and reliability. It features a 32-bit connected architecture, which allows for effortless remote printing. Additionally, it incorporates an advanced extruder mechanism called Nextruder, which ensures accurate filament handling and quick nozzle replacements. A load cell sensor is the core part of its functionality. It automatically ensures a precise initial layer, eliminating the need for manual changes. By integrating Input Shaper support, printing rates are enhanced as it efficiently reduces vibrations, leading to quieter and faster operation. The MK4 prioritizes safety by incorporating built-in power loss recovery and utilizing four thermistors for precise temperature monitoring. It has a build volume measuring 250 × 210 × 220 mm^3. Additionally, the Smooth PEI print sheet ensures excellent adherence and effortless removal of prints. The MK4 enhances the user experience by providing a user-friendly web interface for remote management and monitoring. It also includes a live-view camera for real-time print monitoring and a customizable user interface with simplified one-click printing features. Furthermore, being an open-source platform, it provides advanced users with the ability to customize and experiment. The specifications include a nozzle temperature range reaching 290 °C, a heated bed capable of reaching a temperature of 120 °C, and compatibility with a range of filaments such as PLA, PETG, ABS, and ASA (Original Prusa MK4 3D Printer Kit | Original Prusa 3D Printers Directly from Josef Prusa, n.d.).

4.3 AnkerMake M5 3D Printer

This printer stands out due to its innovative AI-driven high-speed printing mode, which can reach rates of up to 500 mm/s while preserving remarkable detail. This progress results in significantly shorter printing times, demonstrated by the production of a typical Benchy model in just 17 min. The M5's durable aluminum alloy structure and 7 × 7 automatic bed leveling mechanism enables precise printing by providing a consistently even and stable printing surface. The M5 printer has a heated bed that can achieve temperatures up to 100 °C. This allows it to work with a various filaments, such as ABS and PETG. The printer also has an enclosed chamber, which helps regulate temperature and reduces warping, especially for larger prints. In addition to its performance features, the M5 includes user-centric attributes that are designed to streamline the 3D printing process. The AnkerMake app, equipped with an integrated AI camera, actively monitors the progress of printing, sending notifications, and allowing for remote monitoring. In addition, a timelapse feature that is generated automatically provides complete visual documentation of the printing process. The notable specifications of this product include a generous build volume measuring 235 × 235 × 250 mm^3, compatibility with a filament of 1.75 mm diameter, and a nozzle temperature that can reach up to 260 °C (AnkerMake M5 3D Printer—AnkerMake Europe, n.d.; AnkerMake M5 3D Printer—Fast and Precise—Ankermake Europe, n.d.).

4.4 Elegoo Saturn 2 8K Resin Printer

This Printer advances high-resolution printing. It features a 10-inch 8 K monochrome LCD screen, which provides detailed output at a resolution of 7680 × 4320 pixels. Featuring a XY resolution of 28.5 microns, comparable to high-end jewelry printers, each layer solidifies in only 1—3 seconds, hence improving workflow efficiency. This printer has a larger build volume, measuring 219 × 123 × 250 mm^3, which is 43% larger than its previous model. It is equipped with scratch-resistant 9H tempered glass to ensure protection. The primary focus is on ensuring ease of use, which is achieved with a 3.5-inch touchscreen interface and built-in compatibility with Chitubox slicing software. An integrated air filtration system, equipped with an activated carbon filter, effectively reduces resin odors. Additionally, the printer's lifespan is extended using dual heat-conducting copper tubes, which facilitate optimal heat dissipation. The key features of this printer include a Fresnel collimating light source equipped with 64 UV LED lights to ensure precise and polished finishes. It also has a solid Z-axis supported by twin linear rails and a sandblasted build plate surface that provides strong adhesion. Additionally, it is complemented by a PFA release liner film, which makes it easier to remove the printed models (ELEGOO Saturn 2 8k 10" Mono LCD MSLA Resin 3D Printer | Resins3D.Com, n.d.; ELEGOO Saturn 2 Resin 3D Printer, n.d.).

4.5 Anycubic Photon Mono 2

This printer is classified as an innovation in resin 3D printing technology that appeals to users with a larger build volume of 165 × 89 × 143 mm^3 compared to its previous version, it offers more space to create detailed designs. The 6.6-inch 4 K monochrome LCD screen provides high resolution, with an XY resolution of 51 microns. This ensures the creation of prints with complex details, smooth surfaces, and precise edges. In addition, the integration of the enhanced LighTurbo Matrix light source improves the printing speed by promoting even dispersion of light and decreasing the time required for curing to 2 seconds per layer. The user's convenience is improved by several enhancements in the new Photon Workshop 3.0 slicing software, including an intuitive interface and optimized workflows. Additionally, a laser-engraved platform is included to improve print adhesion, and a larger LCD screen protector is provided for increased durability. In addition, the Anycubic Wash and Cure Plus station automates the post-processing duties involved in cleaning and curing finished prints. This results in a more streamlined printing process, making it easier overall (Moura et al.,n.d.).

4.6 Markforged Metal X Gen 2

A metal 3D printer that employs atomic diffusion additive manufacturing (ADAM) technology. It is significantly more cost-effective compared to alternative metal AM methods, offering a cost reduction of up to 10 times. Additionally, it is up to 100 times less expensive than traditional fabrication technologies such as machining or casting. The printer offers a build volume of 300 × 220 × 180 mm^3 and can handle metals and alloys, including stainless steel, tool steel, Inconel, copper, and titanium. The sintering process allows it to attain a layer thickness ranging from 50 to 125 μm. The Metal X Gen 2 incorporates an exterior 7-inch touchscreen, door position sensors, and enhanced chamber insulation, resulting in improved energy and time efficiency compared to its predecessor. Additionally, it provides support for Next Day Metal, a software functionality that enables users to efficiently produce metal components within a period of less than 24 hour through the optimization of debinding and sintering processes. The Metal X Gen 2 features an intuitive interface, a dual-nozzle system, and a cloud-based software platform. Additionally, it can be operated through a mobile application and a tablet device (Mouzakis & Mouzakis, 2018, Natarajan et al., 2020).

4.7 Pam Series MC 3D Printer

The Pam Series MC changes the field of 3D printing by introducing Pellet Additive Manufacturing (PAM) which develops, manufactures, and markets industrial multi-material 3D printers (Original Prusa MK4 3D Printer kit| Original Prusa 3D printers directly from Josef Prusa, n.d.). This technology utilizes pellet feedstock, reducing the limitations associated with conventional powder technologies. Equipped with four extruders, this machine offers flexibility in material selection, allowing for the investigation of metals, ceramics, and thermoplastics in the creation of practical prototypes and small-scale production volumes. Fused Pellet Fabrication (FPE) enables the direct printing of metal and ceramic pellets, which has the potential to decrease expenses and broaden the range of available materials. The MC's industrial capability, featuring a building volume of 300 mm in diameter and 300 mm in height, adeptly manages projects of significant scale. The use of a 40-micron layer resolution and extruder temperatures of up to 450 °C ensures both intricate detail and strength in the final creations, delivering a high level of precision (Pal & Bask, 2022).

4.8 TruPrint 1000 3D Printer

The TruPrint 1000 3D Printer uses laser powder bed fusion (LPBF) technology to fabricate intricate and superior metal components. The printer has a build volume measuring 100 × 100 × 100 mm^3 and can work with metals and alloys, including nickel, cobalt chrome, copper, and titanium. The printer can build at a speed of up to 50 cm^3/hour. The TruPrint 1000 3D Printer is equipped with a multilaser capability, which increases its printing capacity by up to 80% at the same time. It also has a multiplate option that facilitates automatic substrate plate change without interrupting the ongoing build process. Additionally, the printer is equipped with a camera system that provides real-time monitoring of the powder bed and the condition of the printed part. The TruPrint 1000 3D Printer offers user-friendly operation, featuring swift installation, ergonomic manipulation, and an intuitive touchscreen interface. Additionally, it can be operated by a mobile application. The TruPrint 1000 3D Printer is flexible and may be used in a multitude of industries and applications, including industrial, commercial, automotive, aerospace, medical, dental, and musical instruments. It is a highly efficient and cost-effective small-scale metal 3D printer available in the market (Patel et al., 2014).

4.9 EOS M 400 3D Printer

A metal 3D printer that utilizes Direct Metal Laser Sintering (DMLS) technology. The printer has a maximum build capacity of 400 × 400 × 400 mm^3 and can scan at a speed of up to 7 m/s. It can handle metals and alloys, including aluminum, copper, nickel, titanium, and tool steel. The laser has a focus diameter of 90 μm and a power of 1000 W. The EOS M 400 3D printer possesses a modular design, a camera system that observes the building process in real time, and a circulating air filtering technology that reduces maintenance costs (Pollen AM|Pam Series MC, n.d.).

4.9.1 X & L-Series 3D Printers

The X and L-Series 3D printer is a metal 3D printer that uses directed energy deposition (DED) technology to create components with a maximum volume of 1 × 1 × 1 m^3. It can handle metals and alloys, including stainless steel, nickel, titanium, and aluminum, at a deposition rate of up to 7 kg/hr. It features a modular structure and offers many build volumes, allowing for motion in up to 5 axes. It also has a Triple ACT system for dispensing, spreading, and compacting ultra-fine powders with precise parameter control. The X and L-Series 3D printer is equipped with Dynamic Layer Selection technology, which continuously evaluates and adjusts the build quality and accuracy in real-time (Precision CNC Metal Machining—Metal Cutting Corporation, n.d.).

5 Available 3D Printing Materials

This section is devoted to the most common 3D printing material, also currently available in the market and can be purchased directly. The materials are generic and are independent of printer brands.

5.1 Polylactic Acid (PLA)

PLA is a widely employed thermoplastic extracted from environmentally benign resources like corn starch or sugarcane (Quan et al., 2020). The biodegradable polymer is highly favored because of its low melting point and its convenience in FDM printers (Rakshit & Das, 2019). PLA is able to produce items with vibrant colors and a polished surface texture. During the 3D printing process, the PLA filament undergoes heating and is then extruded via a nozzle layer by layer, until the intended object is formed (Alsoufi et al., 2019). PLA offers advantages that go beyond its environmentally friendly origins since it exhibits low warping during the printing process. Although PLA may not possess the same level of heat resistance as many other materials, its flexibility and easy-to-use nature contribute significantly to its widespread use in prototyping, institutions of learning, and consumer applications.

5.2 Polyethylene Terephthalate Glycol (PETG)

Mainly extracted from recycled plastic bottles and food products, it is a thermoplastic polyester made from the PET component (Welle, 2011). The glycol adds durability and strength while also contributing to its high toughness, transparency, and high temperature and chemical resistance (Sepahi et al., 2021).

5.3 Acrylonitrile Butadiene Styrene (ABS)

A highly durable and impact-resistant thermoplastic is widely recognized and used in 3D printing, particularly for creating functional prototypes and end-use components. This petroleum-based polymer shows higher thermal resistance in comparison to PLA, making it appropriate for applications subjected to high temperatures or harsh conditions (Köse Moran, n.d.). The 3D printing procedure for ABS involves raising the temperature of the filament to its melting threshold and forcing it through a nozzle onto a build platform. The ability of ABS to contract during the cooling process may require the use of a heated bed or a closed printing environment in order to reduce warping (Ramian et al., 2021; Zawaski & Williams, 2020). The material's durability, combined with its ability to undergo post-processing methods such as acetone vapor smoothing, enhances the visual and practical attributes of printed objects (Lalehpour et al., 2018).

5.4 Resin

Resin-based 3D printing, exemplified by technologies such as SLA and DLP, presents a distinct approach to the field of additive manufacturing (Dimitrova et al., 2023). Resin printers employ liquid photopolymer resins that undergo solidification upon exposure to light, typically ultraviolet light (Joseph et al., 2022). This technology allows the fabrication of printed objects with great precision and a high-quality surface finish. SLA printers operate by submerging a build platform into a reservoir of liquid resin. The resin layer is cured in a precise manner using either a laser or a light source, resulting in the solidification of the intended shape, done layer by layer. DLP printers employ a digital light projector to rapidly illuminate an entire layer all at once, hence accelerating the printing procedure (Amini et al., 2023).

5.5 Metal

The development of metal 3D printing has become an innovative method for manufacturing complex, durable metal parts with the highest level of precision. Metal 3D printing provides a method to fabricate complex shapes and structures that are unattainable using conventional production techniques. This technique is used in the aerospace, healthcare, and automotive industries, where there is an essential demand for metal components that are both lightweight and efficient (Buchanan & Gardner, 2019). Although the initial cost may be greater in comparison to alternative 3D printing materials, the advantages in terms of design flexibility and material characteristics render metal 3D printing essential for some applications.

5.6 Cellulose

Cellulose presents a promising alternative to non-renewable polymers. Its unique physicochemical properties, which stem from its isolation process, result in distinct chain lengths, levels of crystallinity, and types of hydrogen bonds. These characteristics vary among different types of cellulosic materials. By customizing cellulose and its derivatives, it is possible to develop sustainable inks tailored for efficient and cost-effective 3D printing of objects with specific patterns. Using cellulose-based inks in 3D printing is anticipated to advance Industry 4.0 initiatives, offering diverse applications in fields such as medicine, electronics, and construction (Firmanda et al., 2022).

References

Alsoufi, M. S., Alhazmi, M. W., Suker, D. K., Alghamdi, T. A., Sabbagh, R. A., Felemban, M. A., & Bazuhair, F. K. (2019). Experimental characterization of the influence of nozzle temperature in FDM 3D printed pure PLA and advanced PLA+. *American Journal of Mechanical Engineering*, *7*(2), 45–60. https://doi.org/10.12691/ajme-7-2-1

Amini, A., Guijt, R. M., Themelis, T., De Vos, J., & Eeltink, S. (2023). Recent developments in digital light processing 3D-printing techniques for microfluidic analytical devices. *Journal of Chromatography A*, *1692*, 463842. https://doi.org/10.1016/J.CHROMA.2023.463842

AnkerMake M5 3D Printer—Ankermake Europe. (n.d.). Retrieved February 14, 2024, from https://www.ankermake.com/eu-en/products/m5?variant=43635654164783&discount=WS24M5150

AnkerMake M5 3D Printer—Fast and Precise—Ankermake Europe. (n.d.). Retrieved February 14, 2024, from https://www.ankermake.com/eu-en/m5

Anycubic Photon Mono 2—The First Choice for Resin 3D Printing Beginner—ANYCUBIC-US. (n.d.). Retrieved February 14, 2024, from https://store.anycubic.com/collections/3d-printers/products/photon-mono-2-3d-printer?sscid=21k8_fmofh&

Bambu Lab P1S 3D Printer P1S / In stock|Bambu Lab US. (n.d.). Retrieved February 14, 2024, from https://us.store.bambulab.com/products/p1s?variant=41516337070216&sscid=21k8_fni7c&

Buchanan, C., & Gardner, L. (2019). Metal 3D printing in construction: A review of methods, research, applications, opportunities and challenges. *Engineering Structures*, *180*, 332–348. https://doi.org/10.1016/J.ENGSTRUCT.2018.11.045

Chaudhary, R., Fabbri, P., Leoni, E., Mazzanti, F., Akbari, R., & Antonini, C. (2023). Additive manufacturing by digital light processing: A review. *Progress in Additive Manufacturing*, *8*(2), 331–351. https://doi.org/10.1007/s40964-022-00336-0

Cicala, G., Latteri, A., Curto, B. D., Russo, A. L., Recca, G., & Farè, S. (2017). Engineering thermoplastics for additive manufacturing: A critical perspective with experimental evidence to support functional applications. *Journal of Applied Biomaterials and Functional Materials*, *15*(1), e10–e18. https://doi.org/10.5301/JABFM.5000343/ASSET/IMAGES/LARGE/10.5301_JABFM.5000343-FIG8.JPEG

Daminabo, S. C., Goel, S., Grammatikos, S. A., Nezhad, H. Y., & Thakur, V. K. (2020). Fused deposition modeling-based additive manufacturing (3D printing): Techniques for polymer material systems. *Materials Today Chemistry*, *16*, 100248. https://doi.org/10.1016/J.MTCHEM.2020.100248

Dev Singh, D., Mahender, T., & Raji Reddy, A. (2021). Powder bed fusion process: A brief review. *Materials Today: Proceedings*, *46*, 350–355. https://doi.org/10.1016/J.MATPR.2020.08.415

Dimitrova, M., Vlahova, A., Kalachev, Y., Zlatev, S., Kazakova, R., & Capodiferro, S. (2023). Recent Advances in 3D Printing of Polymers for Application in Prosthodontics. *Polymers 2023*, *15*(23), 4525. https://doi.org/10.3390/POLYM15234525

Dubey, A. K., & Yadava, V. (2008). Laser beam machining—A review. *International Journal of Machine Tools and Manufacture*, *48*(6), 609–628. https://doi.org/10.1016/J.IJMACHTOOLS.2007.10.017

Eladawi, A. E., Gadelmawla, E. S., Elewa, I. M., & Abdel-Shafy, A. A. (2003). An application of computer vision for programming computer numerical control machines *217*(9), 1315–1324. https://doi.org/10.1243/095440503322420241

ELEGOO Saturn 2 8k 10" Mono LCD MSLA Resin 3D Printer|Resins3D.com. (n.d.). Retrieved February 13, 2024, from https://

resins3d.com/products/elegoo-saturn-2-8k-10-mono-lcd-msla-resin-3d-printer

ELEGOO Saturn 2 Resin 3D Printer. (n.d.). Retrieved February 13, 2024, from https://marvle3d.co.nz/scanner-bundle/2143-elegoo-saturn-2-resin-3d-printer.html

EOS M 400—EOS. (n.d.). Retrieved February 17, 2024, from https://na.eos.info/Equipment/Metal-Platforms/EOS-M-400

Fidan, I., Huseynov, O., Ali, M. A., Alkunte, S., Rajeshirke, M., Gupta, A., Hasanov, S., Tantawi, K., Yasa, E., Yilmaz, O., Loy, J., Popov, V., & Sharma, A. (2023). Recent Inventions in Additive Manufacturing: Holistic Review. *Inventions 2023, 8*(4), 103. https://doi.org/10.3390/INVENTIONS8040103

Firmanda, A., Syamsu, K., Sari, Y. W., Cabral, J., Pletzer, D., Mahadik, B., Fisher, J., & Fahma, F. (2022). 3D printed cellulose based product applications. *Materials Chemistry Frontiers, 6*(3), 254–279. https://doi.org/10.1039/D1QM00390A

Folkes, J. (2009). Waterjet—An innovative tool for manufacturing. *Journal of Materials Processing Technology, 209*(20), 6181–6189. https://doi.org/10.1016/J.JMATPROTEC.2009.05.025

FormAlloy X & L Series Machines for 3D Metal Printing. (n.d.). Retrieved February 17, 2024, from https://www.formalloy.com/machines

Ho, K. H., & Newman, S. T. (2003). State of the art electrical discharge machining (EDM). *International Journal of Machine Tools and Manufacture, 43*(13), 1287–1300. https://doi.org/10.1016/S0890-6955(03)00162-7

Jiang, T., Yan, B., Jiang, M., Xu, B., Gao, S., Xu, Y., Yu, Y., Ma, T., & Qin, T. (2023). Study of forming performance and characterization of DLP 3D printed parts. *Materials, 16*(10), 3847. https://doi.org/10.3390/ma16103847

Joseph, B., Sam, R. M., Tharayil, A., Sagarika, V. K., Kalarikkal, N., & Thomas, S. (2022). Photopolymers for 3D printing. *Polymers for 3D Printing: Methods, Properties, and Characteristics*, 145–154. https://doi.org/10.1016/B978-0-12-818311-3.00011-2

Körner, C. (2016). Additive manufacturing of metallic components by selective electron beam melting—A review. *International Materials Reviews, 61*(5), 361–377. https://doi.org/10.1080/09506608.2016.1176289

Köse Moran, M. (n.d.). *Material and Topological Studies for Expanding the use of 3D Printed Parts in Automotive*.

Kumar, K., Kalita, H., Zindani, D., & Davim, J. P. (2019). *Machining Process*. 85–100. https://doi.org/10.1007/978-3-030-21066-3_6

Lalehpour, A., Janeteas, C., & Barari, A. (2018). Surface roughness of FDM parts after post-processing with acetone vapor bath smoothing process. *International Journal of Advanced Manufacturing Technology, 95*(1–4), 1505–1520. https://doi.org/10.1007/S00170-017-1165-5/METRICS

Li, Y., Lee, C. H., & Gao, J. (2015). From computer-aided to intelligent machining: Recent advances in computer numerical control machining research. *229*(7), 1087–1103. https://doi.org/10.1177/0954405414560622

Ligon, S. C., Liska, R., Stampfl, J., Gurr, M., & Mülhaupt, R. (2017). Polymers for 3D printing and customized additive manufacturing. *Chemical Reviews, 117*(15), 10212–10290. https://doi.org/10.1021/ACS.CHEMREV.7B00074/ASSET/IMAGES/MEDIUM/CR-2017-00074G_0037.GIF

Liu, Z., Wang, Y., Wu, B., Cui, C., Guo, Y., & Yan, C. (2019). A critical review of fused deposition modeling 3D printing technology in manufacturing polylactic acid parts. *International Journal of Advanced Manufacturing Technology, 102*(9–12), 2877–2889. https://doi.org/10.1007/S00170-019-03332-X/METRICS

Manufacturing, value added (current US$)|Data. (n.d.). Retrieved February 3, 2024, from https://data.worldbank.org/indicator/NV.IND.MANF.CD

Markforged. (n.d.). *Product Specifications Metal X (Gen 2)*. Retrieved February 17, 2024, from https://s3.amazonaws.com/mf.product.doc.images/Datasheets/Metal_System_Datasheets.pdf

Markforged Announces the Metal X Gen 2, Next Day Metal, and the X7.... (n.d.). Retrieved February 17, 2024, from https://markforged.com/resources/news-events/markforged-announces-metal-x-gen2-next-day-metal-and-x7-field-edition

Martinez Villadiego, K., Arias Tapia, M. J., Useche, J., & Escobar Macías, D. (2021). Thermoplastic starch (TPS)/Polylactic Acid (PLA) blending methodologies: A review. *Journal of Polymers and the Environment, 30*(1), 75–91. https://doi.org/10.1007/S10924-021-02207-1

Moritz, T., & Maleksaeedi, S. (2018). Additive manufacturing of ceramic components. In *Additive manufacturing* (pp. 105–161). Elsevier. https://doi.org/10.1016/B978-0-12-812155-9.00004-9

Moura, A., Fonta, D., & Jorge, H. (n.d.). *Publication review Pollen AM "In the test conducted, the parts showed good shape reproducibility, without structural defects and with a degree of densification in line with expectations. It is therefore considered a technology with potential for application in the technical ceramics segment.» FDM additive manufacturing of ceramic parts: testing PAM technology*.

Mouzakis, D. E., & Mouzakis, D. E. (2018). Advanced technologies in manufacturing 3D-layered structures for defense and aerospace. *Lamination - Theory and Application*. https://doi.org/10.5772/INTECHOPEN.74331

Natarajan, Y., Murugesan, P. K., Mohan, M., Khan, L. A., & S. A. (2020). Abrasive water jet machining process: A state of art of review. *Journal of Manufacturing Processes, 49*, 271–322. https://doi.org/10.1016/J.JMAPRO.2019.11.030

Original Prusa MK4 3D Printer kit|Original Prusa 3D printers directly from Josef Prusa. (n.d.). Retrieved February 14, 2024, from https://www.prusa3d.com/product/original-prusa-mk4-kit-2/#a_aid=CNET

Pal, R., & Basak, A. (2022). Linking powder properties, printing parameters, post-processing methods, and fatigue properties in additive manufacturing of AlSi10Mg. *Alloys, 1*(2), 149–179. https://doi.org/10.3390/alloys1020010

Patel, V. D., Patel2, D. M., Patel, U. J., Patel, B., Butani, N., Scholar, R., & Tech Student, M. (2014). Review of wire-Cut EDM process on titanium alloy. *Journal of Engineering Research and Applications 4*, 112–121. www.ijera.com

Pollen AM|Pam Series MC. (n.d.). Retrieved February 14, 2024, from https://www.pollen.am/pam_series_mc/

Precision CNC Metal Machining—Metal Cutting Corporation. (n.d.). Retrieved May 25, 2024, from https://metalcutting.com/knowledge-center/precision-cnc-machining/

Quan, H., Zhang, T., Xu, H., Luo, S., Nie, J., & Zhu, X. (2020). Photo-curing 3D printing technique and its challenges. *Bioactive Materials, 5*(1), 110–115. https://doi.org/10.1016/J.BIOACTMAT.2019.12.003

Rakshit, R., & Das, A. K. (2019). A review on cutting of industrial ceramic materials. *Precision Engineering, 59*, 90–109. https://doi.org/10.1016/J.PRECISIONENG.2019.05.009

Ramian, J., Ramian, J., & Dziob, D. (2021). Thermal deformations of thermoplast during 3D Printing: Warping in the case of ABS. *Materials, 14*(22), 7070. https://doi.org/10.3390/MA14227070

Rezaeifar, H., & Elbestawi, M. (2022). Porosity formation mitigation in laser powder bed fusion process using a control approach. *Optics & Laser Technology, 147*, 107611. https://doi.org/10.1016/J.OPTLASTEC.2021.107611

Saberi, B. (2018). The role of the automobile industry in the economy of developed countries. *International Robotics & Automation Journal, 4*(3). https://doi.org/10.15406/IRATJ.2018.04.00119

Sepahi, M. T., Abusalma, H., Jovanovic, V., & Eisazadeh, H. (2021). Mechanical properties of 3D-printed parts made of polyethylene terephthalate glycol. *Journal of Materials Engineering and Performance, 30*(9), 6851–6861. https://doi.org/10.1007/s11665-021-06032-4

Srivastava, M., Rathee, S., Patel, V., Kumar, A., & Koppad, P. G. (2022). A review of various materials for additive manufacturing: Recent trends and processing issues. *Journal of Materials Research and Technology, 21*, 2612–2641. https://doi.org/10.1016/J.JMRT.2022.10.015

TruPrint Series 1000|TRUMPF. (n.d.). Retrieved February 14, 2024, from https://www.trumpf.com/en_US/products/machines-systems/additive-production-systems/truprint-1000/?fbclid=IwAR1CvKXRvD1D7Nm0vTTIzUdYj827Z-zIYqERrM1IP4dHAbl46ZBgglpx_jA

Uçak, N., Çiçek, A., & Aslantas, K. (2022). Machinability of 3D printed metallic materials fabricated by selective laser melting and electron beam melting: A review. *Journal of Manufacturing Processes, 80*, 414–457. https://doi.org/10.1016/J.JMAPRO.2022.06.023

UNIDO Statistics Data Portal. (n.d.). Retrieved February 3, 2024, from https://stat.unido.org/database/INDSTAT%202%202023,%20ISIC%20Revision%203

Vladić, G., Banjanin, B., Kašiković, N., & Pavlović, Ž. (2022). Vat photopolymerization. *Polymers for 3D Printing: Methods, Properties, and Characteristics*, 65–74. https://doi.org/10.1016/B978-0-12-818311-3.00018-5

Wang, X., Jiang, M., Zhou, Z., Gou, J., & Hui, D. (2017). 3D printing of polymer matrix composites: A review and prospective. *Composites Part B: Engineering, 110*, 442–458. https://doi.org/10.1016/j.compositesb.2016.11.034

Welle, F. (2011). Twenty years of PET bottle to bottle recycling—An overview. *Resources, Conservation and Recycling, 55*(11), 865–875. https://doi.org/10.1016/J.RESCONREC.2011.04.009

Zawaski, C., & Williams, C. (2020). Design of a low-cost, high-temperature inverted build environment to enable desktop-scale additive manufacturing of performance polymers. *Additive Manufacturing, 33*, 101111. https://doi.org/10.1016/J.ADDMA.2020.101111

Additive Manufacturing of Sustainable Functional Parts

Abdul Hai Alami, Siren Khuri, Fatima Alshamsi, Adnan Alashkar, Ahmad Yasin, Ayman Mdallal, Shamma Alabdulla, Fabrizia Devito, and Mohamad Ayoub

Abstract

The utilization of additive manufacturing as a holistic approach might be the Holy Grail of producing reliable parts while having a low negative impact on the surrounding environment. The production process that requires material and equipment selection can now be predesigned to plan what happens to the part after its useful life. This chapter sheds light on the benefits of additive manufacturing processes and how they seamlessly integrate with modern concepts such as Sustainable Development Goals. Since nature is the genesis of additive manufacturing on land and sea, biomimetics is presented. Finally, rapid investment casting (RIC) as a reliable process for converting polymeric parts into more robust metallic ones is discussed.

A. H. Alami (✉) · F. Alshamsi · A. Yasin · M. Ayoub
Sustainable and Renewable Energy Engineering Department, University of Sharjah, 27272 Sharjah, United Arab Emirates
e-mail: aalalami@sharjah.ac.ae

F. Alshamsi
e-mail: u20200968@sharjah.ac.ae

A. Yasin
e-mail: ahmad.yasin@sharjah.ac.ae

M. Ayoub
e-mail: mohamad.ayoub@sharjah.ac.ae

S. Khuri · A. Mdallal
Sustainable Energy and Power Systems Research Centre, RISE, University of Sharjah, 27272 Sharjah, United Arab Emirates
e-mail: SKhuri@sharjah.ac.ae

A. Mdallal
e-mail: Ayman.Mdallal@sharjah.ac.ae

A. Alashkar
PhD Program in Materials Science and Engineering, American University of Sharjah, 26666 Sharjah, United Arab Emirates
e-mail: b00028197@aus.edu

S. Alabdulla
Mechanical Engineering Department, American University of Sharjah, Sharjah, 26666, UAE
e-mail: g00070854@aus.edu

F. Devito
Department of Mechanical, Mathematics and Management (DMMM), Polytechnical University of Bari, Bari, Italy
e-mail: fabrizia.devito@poliba.it

1 Sustainable Additive Manufacturing Materials

1.1 Conventional Materials

From the functional and sustainable design viewpoint, there exist limitations on suitable materials that can be used for the printing process. The most commonly used materials for additive technologies are polymers (mainly thermoplastics, and to a lesser extent thermosets as they will require post-processing) as well as metals (Jo & Song, 2021). The polymer classes used can be divided into polyamide (PA), polysulphone (PSU), polyetherimide (PEI), polyphenylene sulfide (PPS), polyether-ether-ketone (PEEK), Polyetherketoneketone (Pekk), and bioplastics (PHA). Examples of polymers include polylactic acid (PLA), Acrylonitrile styrene acrylate (ASA), Acrylonitrile butadiene styrene (ABS), Polycarbonate (PC) and glycol-modified polyester (PETG) are already used in commercial settings. In some instances, elastomers are also candidates such as Thermoplastic polyurethane (TPU), Thermoplastic Elastomer (TPE). Polymers have revolutionized additive manufacturing (AM) by offering versatility, durability, and ease of processing. These materials, characterized by their ability to soften when heated and solidify upon cooling, enable intricate designs and complex geometries to be printed one layer at a time. Polymers possess a multitude of properties, from flexibility to high strength, catering to diverse industry sectors ranging from transportation to medicine. Polymers also support sustainable practices in manufacturing by reducing waste through precise material deposition and recycling capabilities. As

A. H. Alami (ed.), *Sustainable Additive Technologies and Innovative Manufacturing Techniques*, Advances in Science, Technology & Innovation, https://doi.org/10.1007/978-3-031-73657-5_5

additive manufacturing continues to evolve, polymers remain at the forefront, driving innovation and expanding possibilities in product design and production.

The utilization of metallic materials in additive manufacturing represents a significant advancement in modern industry, offering unparalleled design freedom, customization, and efficiency. Generally, powdered metals are fused layer by layer, allowing complex geometries and internal structures to be fabricated directly from digital designs. This capability not only reduces material waste but also enables the production of lightweight yet robust components for various industries. The versatility of metallic additive manufacturing extends to creating parts with tailored mechanical properties, corrosion resistance, and heat conductivity, meeting stringent performance requirements that traditional manufacturing methods may struggle to achieve. As technology evolves, the adoption of metallic additive manufacturing continues to grow, promising to redefine how industries approach prototyping, production, and even repair of high-performance metal components.

1.2 Biomaterials

Biomaterials are defined as either natural or synthetic materials that are intended to interact with a biological system. The prospects for the development of AM techniques see a substantial revolution due to the new frontier of biomaterials that are by definition benign to the environment. Bio-fabrication techniques or 3D bioprinting are technologies through which the synthesis of bio-compatible materials from living cells and biomaterials is possible. What was initially shown to be a useful tool for the study and development of regenerative medicine, now also has various applications in the field of research and development of new drugs, and the creation of complex models for studying diseases (Jeong & Atala, 2015).

The basic requirements of any biomaterial are compatibility with living tissue and the possession of all characteristics enabling the material to adequately perform an intended function, such as providing durable support, replacing lost tissue, or promoting the regrowth of damaged tissue. Biocompatible polymers are designed to interact favorably with biological systems without triggering autoimmune responses. Polymeric biomaterials offer significant sustainability advantages due to their versatility and durability and can be easily manufactured in various forms, including solids and viscoelastic materials, making them adaptable for diverse applications. The ability to modulate their chemical, physical, and mechanical properties through the formation of copolymers or blends with different monomers further enhances their functionality. Unlike metals, polymeric biomaterials do not corrode, ensuring longer-lasting performance. Additionally, their density, approximately 1 g/cm^3, is comparable to that of natural fabrics, making them a lightweight and efficient alternative for many uses.

1.3 Composites

Composites are conventionally defined by a polymeric matrix (thermosets or thermoplastics) and a reinforcement (long/continuous fibers, short/chopped fibers, micro-/nanoparticles). Among these materials, the most promising are (short-) fibers reinforced polymer composites (FRPC), as they possess outstanding strength-to-weight ratio. Using composites in additive manufacturing offers a transformative approach to creating high-performance materials with enhanced properties. These composites combine the strengths of multiple materials (mainly polymers as a matrix), resulting in products with superior mechanical, thermal, and electrical characteristics. This customization leads to improved strength-to-weight ratios, increased durability, and tailored performance for specific applications. Moreover, the construction technology inherent in additive manufacturing minimizes waste and allows for the manufacturing of complex geometries.

Most successful applications of FRPC are focused on automotive, energy, and aerospace industries (Goh et al., 2019), where they are typically employed as Carbon Fiber Reinforced Plastics (CFRP), i.e., reinforcements are carbon short fibers. Carbon fiber increases Young's Modulus, and tensile strength, but decreases yield strength, and ductility (Ning et al., 2015). The attractiveness of AM technologies pushed forward the research of a specific family of FRPC and CFRP, where thermoplastics matrix substitutes traditional thermosets, i.e., epoxy resin (Kumar et al., 2018).

2 Metal Recycling and Rapid Investment Casting (RIC)

To produce functional parts, there must be a transition from polymers or composites to metallic materials. As the name implies, the RIC process can be defined as the investment casting (IC) technique, or lost wax process, that replaces an easily malleable pattern with a more durable metallic part. In addition, the efficacy of the manufacturing process in terms of time reduction that the rapid process brings results in a decrease in lead time to the market. Overall, this set of dominant factors plays a major role in the success of RIC where it is currently found at an effective competing level with other foundries that utilize machining. Examples of metallic products developed by this process include flight-critical and safety components for aerospace

Table 1 Challenges for utilizing recycled ferrous and precious metals in rapid investment casting

Recycled ferrous metals	Recycled precious metals
Vulnerable to corrosion/rust due to most metals holding high carbon content	Scarce availability of scrap
Requires intense recycling process (increased tooling cost) characterized by higher energy and resources to cope with the higher melting temperatures of ferrous metals	Higher material cost
Higher production costs and time	Require intense control of the casting process due to its unique properties
Reduced casting quality due to its inconsistent material properties	Requires higher level of purification, i.e., high presence of contaminants
Higher environmental concerns, i.e., carbon dioxide emissions	Requires special handling and processing techniques
Complex supply chain operations	Market dynamics is highly volatile and depends on geopolitical events
More volatile market demand due to its dependence on macroeconomic trends	Requires meticulous quality control measures for applications, i.e., jewelry

applications, X-ray equipment for biomedical applications, dental work, jewelry, and sculptures.

2.1 Utilization of Recycled Metals Through RIC

Since the topic of sustainability became at the forefront of governments and the public's line of importance, and the efficacy of sustainable indicators has become apparent thanks to technological advancements, the current RIC manufacturing industry has shifted towards solutions that synergize between sustainable values and the competitiveness of its principal activities. One of the ways it has created such synergy is by changing the primary practice of casting using virgin metals, which have been generally known to produce robust final parts as a result of their properties, into incorporating recycled metals as a substitute. The range of possible recycled metals that can be used in the casting process is relatively high, consisting of metallic elements that generally fall under the umbrella of ferrous, non-ferrous, and precious metals. With each of these types characterized by its own unique set of properties, a certain recycled metal is selected from this widespread range to produce the final metallic matrix as adhered to the specific requirements for the application. For example, recycled gold holds similar properties to newly mined gold in terms of good quality, color, and texture. Thus, it is the preferred choice for jewelry applications such as creating rings and accessory pieces. While there exist several RIC manufacturers that utilize recycled ferrous and precious metals, the scope of recycled metals that are deemed commonly used in RIC is limited to only non-ferrous metals. This is attributed to the various challenges faced in the recycling of ferrous and precious metals in terms of their technical complexity, material quality, cost considerations, and market demand. Table 1 demonstrates the common challenges in utilizing recycled ferrous and precious metals for rapid investment casting. Non-ferrous metals are characterized by the absence of iron, such as aluminum, copper, titanium, magnesium, zinc, tin, lead, nickel, and their alloys.

2.2 Role of RIC in Achieving Sustainable Development Goals (SDGs)

In 2015, the United Nations developed a set of goals that aim to promote a more sustainable future by tackling global challenges. These goals are dubbed as The Sustainable Development Goals (SDGs) and are outlined in Fig. 1. The importance of utilizing recycled metals in additive manufacturing processes such as the case of RIC, which is in recent times considered a subset of the categories "urban mining" and "circular economy", constitutes a major push force towards achieving the SDGs agenda. According to numerous research published in the literature, the totality of recycled metals in RIC brings a heightened economy and environment through its effective sink under many types of SDGs.

To investigate the role of additive manufacturing in achieving SDGs, an analysis through VOSviewer software was completed linking the total number of keywords relating the subject matter of additive manufacturing with SDGs as shown in Fig. 2. The VOSviewer map provides an insightful visual representation of how additive manufacturing contributes to global sustainability efforts. The map illustrates clusters of keywords that reveal the interplay between various aspects of 3D printing and SDGs. For instance, one cluster connects "additive manufacturing," to "sustainable production," thus highlighting the role of 3D printing in achieving SDGs. Larger nodes represent more commonly

Fig. 1 Sustainable development goals (sustainable development goals—Wikimedia commons, n. d.)

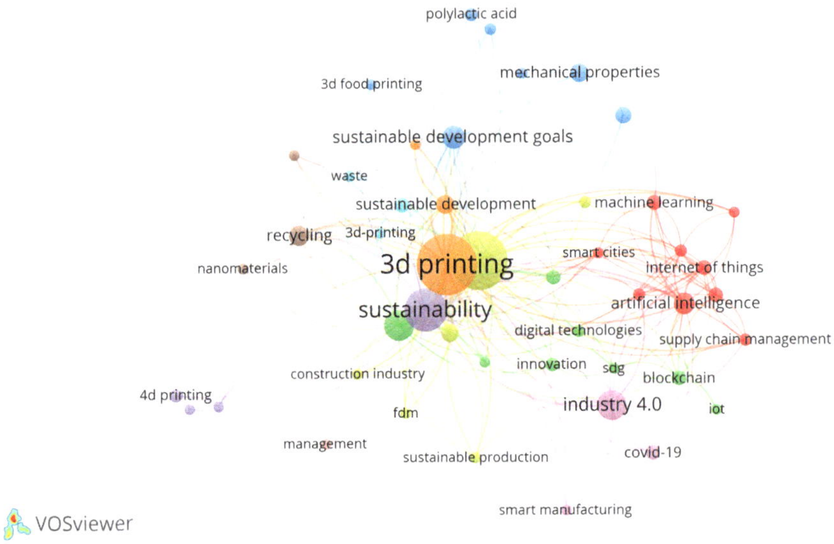

Fig. 2 VOSviewer analysis of AM technology with SDGs

used keywords, signifying their prominence in the research. Hence, the connection between 3D printing, sustainability, and circular economy can be clearly seen and emphasizes the role of additive manufacturing in driving progress toward sustainability, uncovering potential synergies between technological advancements and sustainable development initiatives, and guiding future research and policy directions to maximize the contributions to SDGs.

The SDGs that can be achieved through the utilization of recycled metal and RIC can be restrained to SDGs 6, 7, and 8, and the impact of recycled metal in additive manufacturing is prominent.

The incorporation of sustainable management to any kind of water resource serves as the epitome of SDG 6. The dependence on recycled metals as the source in the RIC process helps increase the strive towards SDG 6, which emphasizes ensuring clean and sanitation goals of water resources for all. Through the potential of recycling metals from waste, the inherent environmental disadvantage associated with landfill disposal and accumulation gets eradicated. Simply its ability to reduce and remove the waste from landfills entails a decrease in the possibility of contaminating water bodies which is often witnessed by the landfill's release of toxic compounds. It is important to note that this method of saving water resources from contamination can lead to enhanced water quality while serving the purpose of creating economic opportunities.

The alignment of SDG 7 in relation to utilizing recycled metals in the RIC process has an overall indirect impact where this formidable alternative to the line of conventional extraction of virgin metal ores allows for less energy consumption than the latter. This drastic reduction shift in energy intensity helps to enhance the energy efficiency, which represents one of the key goals under SDG 7 (Tier 7.3).

The alignment of utilizing recycled metals in manufacturing processes such as RIC for meeting SDG 8 lies predominantly under the category of economic growth. There exist four factors which include job creation, supply chain resilience, cost savings, and market expansion that are at the forefront in contributing to economic growth for the use of recycled metals. Similar to extracting primary metals from ores, employment opportunities are created for such utilization of recycled metals due to its procurement from the recycling process that generally involves extensive steps that include collection, sorting, processing, and manufacturing. These opportunities have significantly expanded in the manufacturing sectors as a result of the high levels of demand for recycled metals that are now in place and hence raises the market value overall.

3 Benefits of Utilizing Recycled Metals in RIC

3.1 Technical Benefits

Similar to primary metals, recycled metals share excellent characteristics such as high tensile and yield strengths as well as ductility. With the current advancements in recycling technologies and processes, the end result of the recovery cycle constitutes a metallic-based product with minimized impurities, allowing it to exhibit relatively better performance than its primary counterpart. Whether holding in par or even higher than those of primary metals, recycled metals are not limited to strength and ductility but also share other important parameters that serve as technical requirements for RIC such as consistency, flexibility, versatility, high corrosion resistance, fatigue resistance, dimensional accuracy, and intricate detailing. In addition to excellent surface finishes, thermal stability, and high purity. These properties help ensure suitability for precision casting applications, utilization for a wide range of casting applications, desired mechanical and physical properties, longevity, and reliability. Moreover, they allow for the reduction in the need for additional machining and finishing operations and provide good performance under varying temperature conditions without significant degradation, leading to an overall improved quality with fewer casting defects, respectively.

3.2 Environmental Benefits and Mitigation of Health and Safety Risks

The utilization of recycled metals eliminates the need for using primary metals obtained from mining their ores, thus eradicating the environmental disadvantages associated with the process. The opportunities created by utilizing recycled metals for RIC include the conservation of natural resources and the reduction of environmental impact.

One of the main challenges with consistent and continuous virgin metal extraction via mining operations involves the scarce availability of the ore grades and its declining rate. Contrary to their virgin counterparts, the availability of recycled metals provides a steady and unrestricted flow whether it originates from end-of-life products (i.e., electronic waste) or from scrap streams produced during manufacturing processes. Through utilizing recycled metals in RIC, the natural and valuable metal ores remain saved from thorough depletion and as a result provide a source of economic value for future generations' needs, if all necessary alternatives such as recycling fail (Söderholm & Ekvall, 2020).

The role of ore mining for virgin metal extraction plays a negatively substantial impact on the environment. A major well-known argument for its consideration as an inefficient process from an environmental perspective lies in its link to direct and indirect heavy concentrations of greenhouse gas released into the atmosphere. The release of indirect harmful greenhouse gasses is not limited to only the extracting process but can also be found linked to other various activities, technologies, and auxiliary equipment involved in the whole mining and transportation processes. Meanwhile, the high direct greenhouse gas emissions are credited to the extremely high energy consumption that is generally required to electrify the mining operations. Moreover, with the projected decrease in the metal content of metallic ores that has been increasing lately, and in parallelism with the continuous use of mining, an even greater consumption of energy will be favored to achieve optimum required concentrate grades and thus enable an associated further increase in direct greenhouse gasses emissions derived from the reliance of fossil fuels as the beneficiary energy source. The consequence of this atmospheric contamination not only brings about a disruption in climate and a faster tendency towards global warming, but also carries with it a high-cost liability in terms of environmental damages, economic, and social impact.

3.3 Social Benefits

Through minimizing the functionality of refining primary metals, the employment opportunities created for such conventional method of extraction will not go to waste as the recycling industry jobs specifically the processes involved in the collection, sorting, processing, refining, and quality control of scrap metals create an entirely new diverse and stable job market. Thereby contributing to supporting communities and reducing unemployment rates. With the current state of demand for recycled metals reaching high levels, such opportunities have substantially increased, which in turn increases the overall market value.

The practice of metal recycling sets the process following the model description of "Waste to Wealth". This streamlined pathway aligns with converting end-of-life metallic products from landfills into useful new products allowing for the loop closure of metal supply chains. This ability to establish a sustainable cyclic process and disrupt the traditional linear model of metal extraction, which typically ends at disposal, aligns with the principles of a circular economy. These principles emphasize 'recycling' and 'reusing' to create an additional process stream, effectively returning materials to the initial stage of raw material production within the circular management cycle.

3.4 Economic Benefits

By leveraging recycled materials, RIC manufacturing processes are positioned as economically viable due to the quantifiable cost savings achieved in comparison to the extraction of newly mined metals. The totality of these cost savings arises from a variety of cost indicators involved in the overall lifecycle of the recycling process, which includes raw materials, energy, environmental, waste management, and operational costs. The specific potential of each cost structure translates to lower and stable purchasing prices, reduced utility bills and operational costs, lower waste disposal, management, maintenance, and transportation costs, a reduction in RIC's carbon tax liability, and higher revenue and long-term financial gains in comparison to the value of metallic ore extraction. The basis for such economic gains is attributed to the environmental, technical, and social benefits of recycled metals as well as the importance of the recycled metal market, which has witnessed an increase in demand in both developed and developing countries. Depending on the type of recycled metal, the economic savings can range between 50 and 90% for processes associated with aluminum production (Damanhuri, 2012). The marginal benefit of recycling 1 ton of metallic material holds significant cost savings that vary depending on the country and region where the recycling occurs.

4 Biomimetic Additive Manufacturing

4.1 Role of Biomimetics in Process Design and Materials Selection

Additive manufacturing is a technique that is heavily inspired by nature's intricate designs and structures, such as beehives, spider webs, and various shell structures. These natural architectures are the outcome of millennia of evolutionary optimization (Wenzel, 1993). Table 2 shows examples of structures built by various living organisms that are convenient to study for further development in the additive manufacturing research field.

When comparing Western Honeybee and Tidarren Sisyphoides Spider, it is determined that both organisms utilize biogenic materials, with bees using wax for constructing hexagonal lattices in hives, and the spider using silk fibers for cobwebs. Bees methodically add wax to construct durable, gravity-defying structures. In contrast, spiders quickly establish a primary structure followed by gradual reinforcement, creating a tangled web pattern that efficiently absorbs impacts from flying prey. Beehives demonstrate exceptional stability relying solely on the geometric strength of hexagonal lattice (Designed Research & Performed Research, 2021), While Spider webs increase in strength and toughness with the density of silk, showcasing adaptability and repairability (Franklin et al., 2022a, 2022b).

Mollusks Shellfishes and Armadillos' tough outer structures may resemble one another as hard shells however, the materials used in each of those structures are drastically different. Mollusks build shells from bio-composites on a protein plate, whereas armadillos have bony plates made from keratin. Both structures serve protective functions for the living organism within. Mollusk shells are mainly defensive against predators and environmental factors, while armadillo osteoderms provide armor-like protection and flexibility due to embedded collagen fibers. The properties of mollusk shells are highly influenced by water salinity, affecting shell thickness and strength. In contrast, armadillo plates develop post-birth, with flexibility built into the structure without significantly compromising its strength (Clark et al., 2020).

Allomyrina Dichotoma Beetles and Sea Urchins both use biological materials but in different configurations. Beetles have chitin fibers and proteins forming a sandwich-like microstructure in forewings, while sea urchins use a cellular material to form a complex, porous skeletal structure. Beetle forewings show a complex internal microstructure designed for strength with load-bearing capabilities enhanced by trabeculae. Sea urchin spines exhibit a gradient in porosity which enhances strength, with thicker and shorter branches at certain segments. In beetles, the integration of trabeculae significantly increases interlaminar shear strength, whereas in sea urchins, the porosity gradient in spines optimizes structural robustness (Zhou et al., 2020).

4.2 Learning from Nature's Architects

Biomimetic additive manufacturing is an approach that is heavily based on nature's design principles to create highly efficient structures. The greatest example of this is the

Table 2 A summary of compatible structures by living organisms suitable for AM

Organism/species	Structure	Material	Method	Pattern	Time	Remarks	Strength	References
Western honey bee (Apis mellifera)	Honey bee hives (Honey cells)	Wax	Within two hours, bees significantly augment the edges of the comb's narrow sections in the newer, lower cells. The older cells at the top of the comb become deeper as the bees focus on both deepening existing cells and constructing new ones at the leading edge of growth	Hexagonal grids arranged staggered on either side of the comb's central spine	28 h	Depth of the cell varies depending on its location	Exceptional stability against gravitational force	Franklin, Niverty, Harpur, Chawla, et al. (2022a, 2022b)
Tidarren sisyphoides spider	Cobweb	Silk fibers	Development strategy comprises of prioritizing a quick construct of the structure taken after by moderate structural reinforcement while utilizing silk as fabric and keeping the structure stable and functional all through development. This approach makes a structure that is steady, useful, repairable, and continuously reinforceable	Tangle web which is a stretchable Primary structure that is attached to frame supports	The primary structure is built within 2 days, at that point, the structure is strengthened without any expanding extensions over the next couple of days	Efficiently absorbs projectile impacts (prey flying into cobweb)	Web strength and toughness increases with the increase in density of silk fibers	Su et al. (2021)
Mollusc shellfish —mussels, scallops, clams, oysters	Mollusc shell	Bio-composites and organic substrate	The basic shell-making process is shared among shellfish, in which the fleshy shell exudes a calcareous crust onto a protein plate called periostracum	Hierarchical bio-composites where the hard mineralized phase is embedded in a softer organic substrate	Varies for every species	Highly dependent on region and genetics	Water Salinity highly influences the thickness and strength of the hard shell	Clark et al. (2020)

(continued)

Table 2 (continued)

Organism/species	Structure	Material	Method	Pattern	Time	Remarks	Strength	References
Armadillo	Osteoderm	Boney plates made from keratin	Osteoderm is built on top of the skin layer of invertebrates and is attached to it in the form of individual plates	Collagen fibers are attached to the more rigid parts of the plates which increases flexibility without negatively affecting strength	6 weeks	Built on top of the skin layer	Flexibility due to structure does not significantly affect the strength of the osteoderm structure	Maden et al. (2023)
Allomyrina dichotoma Beetle	Forewings	Chitin fibers and proteins	Develops in the pupa stage as beetles experience a complete metamorphosis cycle	The internal structure of the forewings features a sandwich-like configuration supported by trabeculae. It includes layers of unequal thickness in the upper and lower skin, as well as a honeycomb core with trabeculae	Two months	The addition of trabeculae increased the elastic buckling critical load of the honeycomb cell model by 41% compared to the model without trabeculae	Trabeculae contributed significantly to enhancing the interlaminar shear strength by bearing a substantial portion of the interfacial shear force	Zhou et al. (2020)
Sea urchin	Sea urchin spine	Biological cellular material	The cellular structure comprises a network of branches connected at nodes, forming a functional skeleton	Complex porous structure	Skeletal structure was studied after adult sea urchin was deceased	The stereo structure shows a controlled gradient in porosity	Branches along chains are thick and short enhancing the strength of the structure	Yang et al. (2020)
Silkworm	Cocoon	Silk fibers	Silk made primarily from a protein called fibroin is secreted from silk glands and spun over a period of time to form a cocoon structure	Multiple layers loop around to form the cocoon structure	Observation done over a long period of time	Energy conservation during the silk-spinning process was observed highlighting the importance of mimicking this process in artificial systems	Can withstand up to 35% elongation with very high tensile strength when compared with high-performance synthetic fiber	Kaplan et al. (1993)
Beaver	Beaver dam	Sediments, wood, or both sediments and wood (a dominant dam material was assumed to be constant for this study)	Materials are layered depending on the composition of the dam. The dam offers a hollow space underneath it to shelter beavers from predators and harsh winter conditions while still maintaining a strong build structure	Groundwater-fed dams contain more sediment layers while stream-fed dams contain more wood layers	Daily observations were made over a long period of time to study both low and high-elevation pond	Material highly affected water flow rate, type, and water storage capacity	Structure and strength were highly dependent on the material it was made out of but it is worth mentioning that all of them can withstand high water flow rates and types	Ronnquist and Westbrook (2021)

(continued)

Table 2 (continued)

Organism/species	Structure	Material	Method	Pattern	Time	Remarks	Strength	References
Termite	Termite nest	Materials vary from one species to another, some have paper-like nests and others have cement-like mounds	The external structure is built out of a stronger material (paper-like or cement-like) The internal structure is made of materials that are not as strong (fecal matter mixed with saliva or soil)	Materials are all reinforced with saliva and compressed to form a dense yet porous material	A nest was observed for an unspecified long period of time	Nesting structures provide buffering to environmental changes such as fluctuations in temperature, ventilation, and humidity	Strength depends on the material therefore varies however it has exhibited antifungal and antimicrobial properties	Li et al. (2022)
Gecko (Tokay gecko)	Foot sole structure	Geckos lack glandular tissues on their toes thus no adhesive substances are secreted the structure relies solely on its mechanical design	Geckos adhere to surfaces using van der Waals interactions alone. Toe pads are covered by rows of setae (a bristle-like structure) that form the attraction with the surface	Toe pads are covered uniformly in rows of setae. Each strand of seta ends with a frayed-looking end that consists of a large amount of spatulae	The structure slides, attaches, and detaches in a matter of seconds with minimal force	Effortless detachments that are rapidly repeated as the gecko walks on surfaces are arguably more important to study as their capabilities are not available in synthetic adhesives	The structure generates parallel frictional force equal to or greater than body weight	Autumn and Peattie (2002)
Chameleo Chameleon	Chameleon skin	The main component is made out of chromatophores with a combination of melanophores	Scales of varying sizes are uniformly arranged in rows, overlapping each other. The scales are wider than they are tall, and the entire surface displays micro-ridges. Deep ridges separate each scale from the next. Each scale features a prominent lenticular epidermal sensory organ	Conical or elliptical-shaped structures, with scales of varied sizes	The study observed juvenile chameleons throughout the summer of 2015	In response to external stimuli, chameleon skin changes color and properties	The outer skin surface is thin, flexible, and made out of 5 major layers	Fouda and El Mansi (2017)

incorporation of wasp nests in 3D printing. Inspecting paper wasp nests that are constructed by social wasp colonies, shows an exceptional combination of strength and flexibility while managing to maintain a lightweight architectural structure. These structures are the result of complicated biological processes that show the intelligence and adaptability of the wasps in response to environmental challenges (Wenzel, 2020). Mimicking these nature-inspired designs in additive manufacturing allows the creation of materials and structures with enhanced properties.

One of the most interesting aspects of wasp nest-inspired additive manufacturing is the use of optimized geometries. Through generations of evolution, wasps have perfected the art of constructing nests with the most efficient use of resources not to mention the impressive self-organizing behavior of wasp colonies that serves as a model for decentralized and adaptive manufacturing processes (Smith et al., 2023; Wenzel, 1993). Additive manufacturing processes can replicate these optimized geometries, resulting in lightweight structures that maintain structural integrity for long periods of time. This approach is particularly beneficial in the transportation industries, where weight heavily impacts fuel efficiency and performance. In additive manufacturing, this principle can be applied to a construction method developed so that it can adapt to changing conditions during the printing process. This flexibility allows for the production of complex structures with minimal material waste.

4.3 The Future of Manufacturing: Wasp Nest-Inspired Biomimetic Additive Manufacturing

It is evident that wasp nest-inspired designs offer unique solutions to some of the challenges faced by traditional manufacturing processes. The complex architectural design of wasp nests which are characterized by hexagonal cells woven with a high degree of precision, has inspired innovative 3D printing technologies. Researchers and engineers are continuously aiming to enhance the strength-to-weight ratio of printed materials, making the hexagonal structure ideal for applications where both durability and lightness are critical (Kunzmann et al., 2023). This biomimetic approach not only improves efficiency but also allows for decentralized manufacturing without the use of detachable supports to hold the structure during the building process; thus, reducing wasted material and decreasing the reliance on centralized production methods. The environmental sustainability of wasp nest-inspired biomimetic additive manufacturing is another key point to mention. By using optimized geometries and minimizing material waste, this approach aligns with the principles of eco-friendly manufacturing (Jiang et al., 2018).

Wasps exhibit exceptional intelligence and precision when building their nest. The wasp begins by selecting suitable materials, such as chewed wood or plant fibers, which it mixes with saliva to create a malleable substance (Perveen & Shah, 2013). Utilizing their mandibles, wasps construct hexagonal cells in a symmetrical pattern, optimizing space and material usage. The wasp's ability to adapt its construction based on environmental conditions and available resources is a testament to its decentralized decision-making and adaptive capabilities (Wenzel, 2020). This process mirrors the efficiency of 3D printing technology in many ways. Just like a wasp adds layers to its nest, a 3D printer builds an object layer by layer, following a predetermined digital design. Both processes emphasize precision, resource optimization, and the ability to adapt to changing conditions, although a 3D printer operates under human programming rather than innate instinct. Both nature and technology showcase the power of systematic construction, emphasizing the potential for biomimetic additive manufacturing to take inspiration from the ingenious processes observed in the natural world.

4.4 Bio-inspired Hexagonal Structure: A Case Study

Gaining inspiration from nature's designs, a 3D hexagonal structure that mimics a real wasps nest's dimension ratio was designed to investigate its mechanical performance, specifically under compression. Figure 3 shows the structure specifically designed for the test. This design was set to be a single-walled hexagonal pattern with equal sides (6 mm), fixed inner angles (120 degrees), and a thickness (0.1 mm) that is proportional to the hexagonal side's length.

The sample was printed via a (Flash Forge Creator 3 Pro) 3D printer with a 0.4 mm nozzle setting which was the predominant factor in controlling the thickness of the walls. It is also relevant to mention the material of choice which was set to be PLA. Figure 4 shows the final printed sample prior to and after the compression test. When observing the honeycomb structure after the compression test has been done, it appears that some of the cell walls are bent or collapsed, indicating local instability and failure happening under this compression test. The overall deformation seems to be somewhat uniform across the structure, suggesting that the load was fairly evenly distributed across the honeycomb. However, some cells appear more deformed than others, which might indicate localized weaknesses or variability in the structure. Looking at the sample after compression, it seems like the deformation started at the top right side of the

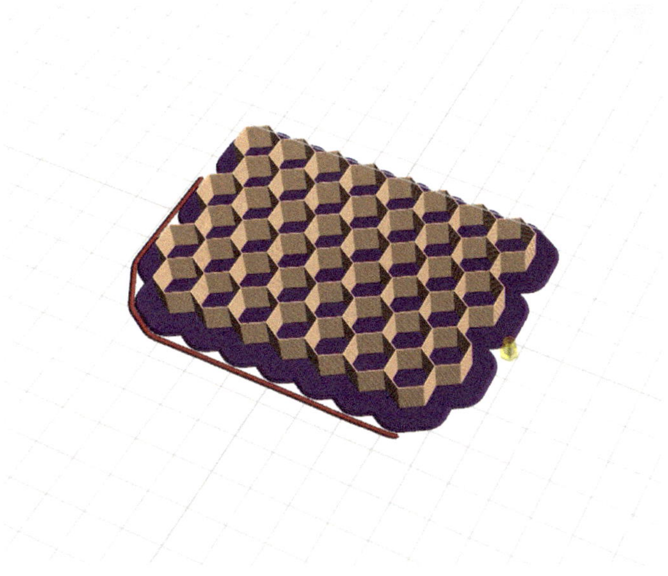

Fig. 3 CAD model of the hexagonal pattern

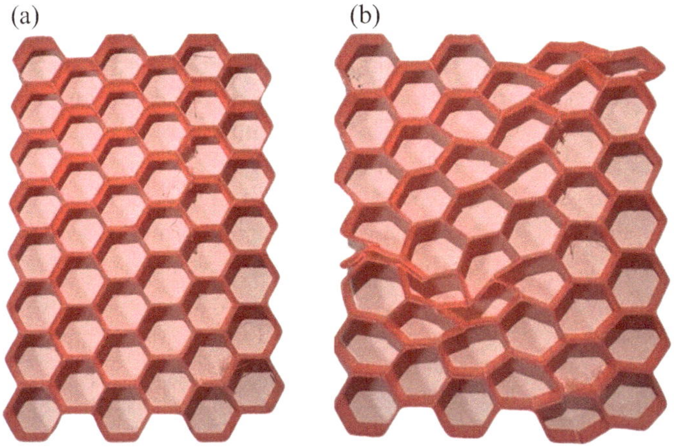

Fig. 4 3D printed hexagonal pattern **a** before and **b** after compression test

sample and followed a diagonal path all the way to the middle of the sample. This path shifted back diagonally to the left as the deformation reached the lower cells of the sample.

Figure 5 shows the elastic performance of the 3D-printed hexagonal honeycomb. It seems that the structure has undergone significant compression before reaching a point of instability. The curve's fluctuations suggest ongoing buckling and collapse of cell walls during the test. The initial steep part of the curve indicates that the pattern is initially uniform, but the later part of the curve shows fluctuations that indicate non-uniform deformation, likely due to the observed progressive buckling and collapse of individual cells. The initial steep slope of the graph, followed by a series of peaks, suggests a series of yielding and buckling events and is indicative of a high elastic modulus. The initial slope can be used to estimate the elastic modulus, the peak points can represent the yield strength and the highest point before a major drop can be considered as the ultimate strength. The material also shows some ductility as evidenced by its ability to undergo significant strain before failure. However, the abrupt drops in load capacity at certain points suggest brittle failure modes at local points within the sample possibly due to the hexagonal structure. The area under a normal force vs strain curve represents the energy absorbed by the structure. This determines the quality of the energy absorption of the structure before failure. The exact energy absorbed can be calculated through the integration of the stress–strain curve.

The primary failure indicators observed are the buckling of cell walls and the complete collapse of some cells. These contribute to the structure's loss of load-bearing capacity as deformation progresses. The breaking of some cell wall sides was also observed. However, no snapping sounds were heard which indicates ductile fractures due to weak points in the structure. Parameters such as the ratio of layer height to hexagonal side length, print speed, and nozzle temperature critically affect the layer adhesion and material properties in

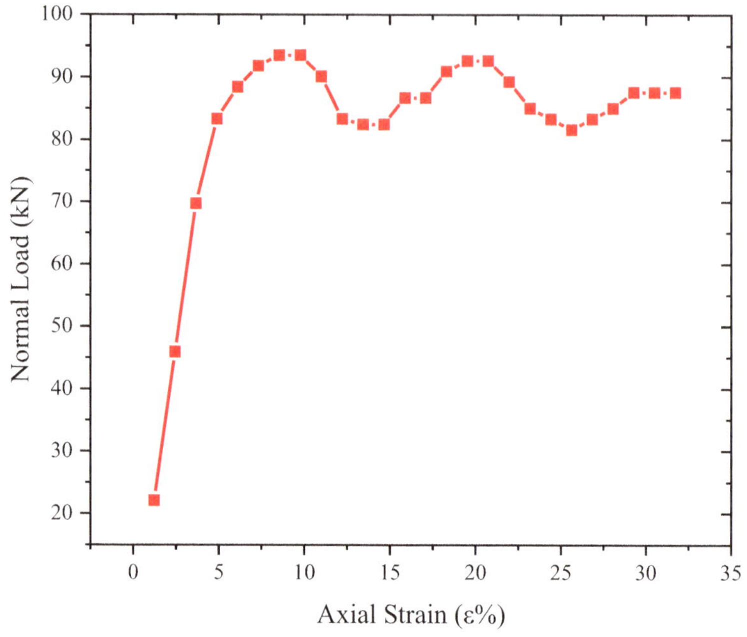

Fig. 5 Stress–strain curve of the 3D printed hexagonal pattern

3D printing. Adjusting the nozzle temperature to ensure smoother material flow and better adhesion, and optimizing print speed and layer height for uniformity could enhance structural integrity (Alsoufi et al., 2019).

The peak of the normal load vs Axial strain appears to be around 94 kN. The weight of the Sample determined by its mass is 35.32 N. This allows for the calculation of the strength-to-weight ratio (α) using the following formula (Pernet et al., 2022):

$$\alpha = \frac{\text{maximum Load}}{\text{Weight of sample}} \quad (1)$$

The calculated strength-to-weight ratio obtained is around 2660, which falls in an acceptable region when compared to carbon fiber (2457) and Kevlar (2514) (Singh, 2021).

References

Autumn, K., & Peattie, A. M. (2002). Mechanisms of adhesion in Geckos. *Integrative and Comparative Biology, 42*(6), 1081–1090. https://doi.org/10.1093/ICB/42.6.1081

Clark, M. S., Peck, L. S., Arivalagan, J., Backeljau, T., Berland, S., Cardoso, J. C. R., Caurcel, C., Chapelle, G., De Noia, M., Dupont, S., Gharbi, K., Hoffman, J. I., Last, K. S., Marie, A., Melzner, F., Michalek, K., Morris, J., Power, D. M., Ramesh, K., Harper, E. M., et al. (2020). Deciphering mollusc shell production: the roles of genetic mechanisms through to ecology, aquaculture and biomimetics. *Biological Reviews, 95*(6), 1812–1837. https://doi.org/10.1111/BRV.12640

Damanhuri, E. (2012). *Post-consumer waste recycling and optimal production Edited by Enri Damanhuri. May*. https://books.google.com/books/about/Post_Consumer_Waste_Recycling_and_Optima.html?id=ueKdDwAAQBAJ

Designed Research; I, M. J. B. S., & Performed Research; R, M. J. B. M. (2021). *Contributed new reagents/analytic tools, 118*(33). https://doi.org/10.1073/pnas.2101296118/-/DCSupplemental

Fouda, Y. A., & El Mansi, A. A. (2017). Skin pattern structure and function of juvenile ages of Chameleo chameleon. *Egyptian Journal of Basic and Applied Sciences, 4*(1), 22–29. https://doi.org/10.1016/J.EJBAS.2017.01.002

Franklin, R., Niverty, S., Harpur, B. A., & Chawla, N. (2022). Unraveling the Mechanisms of the Apis mellifera Honeycomb Construction by 4D X-ray Microscopy. *Advanced Materials, 34*(42). https://doi.org/10.1002/adma.202202361

Franklin, R., Niverty, S., Harpur, B. A., Chawla, N., Franklin, R., Niverty, S., Chawla, N., & Harpur, B. A. (2022). Unraveling the mechanisms of the Apis mellifera Honeycomb construction by 4D X-ray microscopy. *Advanced Materials, 34*(42), 2202361. https://doi.org/10.1002/ADMA.202202361

Goh, G. D., Yap, Y. L., Agarwala, S., & Yeong, W. Y. (2019). Recent progress in additive manufacturing of fiber reinforced polymer composite. *Advanced Materials Technologies, 4*(1). https://doi.org/10.1002/admt.201800271

Jeong, C. G., & Atala, A. (2015). *3D printing and biofabrication for load bearing tissue engineering* (pp. 3–14). https://doi.org/10.1007/978-3-319-22345-2_1

Jiang, J., Xu, X., & Stringer, J. (2018). Support structures for additive manufacturing: A review. *Journal of Manufacturing and Materials Processing, 2*(4). MDPI. https://doi.org/10.3390/jmmp2040064

Jo, B. W., & Song, C. S. (2021). Thermoplastics and photopolymer desktop 3D printing system selection criteria based on technical specifications and performances for instructional applications. *Technologies, 9*(4), 91. https://doi.org/10.3390/technologies9040091

Kaplan, D., Adams, W. W., Farmer, B., & Viney, C. (1993). Silk: Biology. *Structure, Properties, and Genetics, 16*, 2–16. https://doi.org/10.1021/BK-1994-0544.CH001

Kumar, D., Rajmohan, T., & Venkatachalapathi, S. (2018). Wear behavior of PEEK matrix composites: A review. *Materials Today: Proceedings, 5*(6), 14583–14589. https://doi.org/10.1016/j.matpr.2018.03.049

Kunzmann, C., Aliakbarpour, H., & Ramezani, M. (2023). Biomimetics design of sandwich-structured composites. *Journal of Composites Science 2023, 7*(8), 315. https://doi.org/10.3390/JCS7080315

Li, H., Greening, C., Li, H., & Greening, C. (2022). Termite-engineered microbial communities of termite nest structures: A new dimension to the extended phenotype. *FEMS Microbiology Reviews*, *46*(6), 1–13. https://doi.org/10.1093/FEMSRE/FUAC034

Maden, M., Polvadore, T., Polanco, A., Barbazuk, W. B., & Stanley, E. (2023). Osteoderms in a mammal the spiny mouse Acomys and the independent evolution of dermal armor. *ISCIENCE, 26*, 106779. https://doi.org/10.1016/j.isci.2023.106779

Ning, F., Cong, W., Wei, J., Wang, S., & Zhang, M. (2015, June 8). Additive manufacturing of CFRP composites using fused deposition modeling: Effects of carbon fiber content and length, *vol 1. Processing*. https://doi.org/10.1115/MSEC2015-9436

Pernet, B., Nagel, J. K., & Zhang, H. (2022). Compressive strength assessment of 3D printing infill patterns. *Procedia CIRP, 105*, 682–687. https://doi.org/10.1016/j.procir.2022.02.114

Perveen, F., & Shah, M. (2013). Nest architectural patterns by three wasp species (Vespa velutina, Polistes flavus and Sceliphron formosum) with reference to their behavior . *International Journal of Insect Science*, *5*, IJIS.S10737. https://doi.org/10.4137/ijis.s10737

Ronnquist, A. L., & Westbrook, C. J. (2021). Beaver dams: How structure, flow state, and landscape setting regulate water storage and release. *Science of the Total Environment, 785*, 147333. https://doi.org/10.1016/J.SCITOTENV.2021.147333

Alsoufi, S. M., Alhazmi, W. M., Suker, K. D., Alghamdi, A. T., Sabbagh, A. R., Felemban, A. M., & Bazuhair, K. F. (2019). Experimental characterization of the influence of nozzle temperature in FDM 3D printed pure PLA and advanced PLA+. *American Journal of Mechanical Engineering*, *7*(2), 45–60. https://doi.org/10.12691/ajme-7-2-1

Singh, D. K. (2021). Strength of materials. *Springer International Publishing*. https://doi.org/10.1007/978-3-030-59667-5

Smith, M. L., Loope, K. J., Chuttong, B., Dobelmann, J., Makinson, J. C., Saga, T., Petersen, K. H., & Napp, N. (2023). Honey bees and social wasps reach convergent architectural solutions to nest-building problems. *PLoS Biology*, *21*(7 July). https://doi.org/10.1371/journal.pbio.3002211

Söderholm, P., & Ekvall, T. (2020). Metal markets and recycling policies: Impacts and challenges. *Mineral Economics, 33*(1–2), 257–272. https://doi.org/10.1007/S13563-019-00184-5/FIGURES/4

Su, I., Narayanan, N., Logrono, M. A., Guo, K., Bisshop, A., Mühlethaler, R., Saraceno, T., & Buehler, M. J. (2021). In situ three-dimensional spider web construction and mechanics. *Proceedings of the National Academy of Sciences of the United States of America, 118*(33), e2101296118. https://doi.org/10.1073/PNAS.2101296118/SUPPL_FILE/PNAS.2101296118.SM01.MP4

Sustainable Development Goals—Wikimedia Commons. (n.d.). Retrieved July 24, 2024, from https://commons.wikimedia.org/wiki/File:Sustainable_Development_Goals.svg

Wenzel, J. W. (1993). Application of the biogenetic law to behavioral ontogeny: A test using nest architecture in paper wasps. *Journal of Evolutionary Biology, 6*(2), 229–247. https://doi.org/10.1046/j.1420-9101.1993.6020229.x

Wenzel, J. W. (2020). Nest structure: Social wasps. In *Encyclopedia of social insects* (pp. 1–14). Springer International Publishing. https://doi.org/10.1007/978-3-319-90306-4_146-1

Yang, T., Wu, Z., Chen, H., Zhu, Y., & Li, L. (2020). Quantitative 3D structural analysis of the cellular microstructure of sea urchin spines (I): Methodology. *Acta Biomaterialia, 107*, 204–217. https://doi.org/10.1016/J.ACTBIO.2020.02.034

Zhou, M., Huang, D., Su, X., Zhong, J., Hassanein, M. F., & An, L. (2020). Analysis of microstructure characteristics and mechanical properties of beetle forewings, Allomyrina dichotoma. *Materials Science and Engineering: C, 107*, 110317. https://doi.org/10.1016/J.MSEC.2019.110317

Recent Advances in Additive Manufacturing Technologies

Abdul Hai Alami, Haya Aljaghoub, Mohamad Ayoub, Adnan Alashkar, Ahmad Yasin, and Ayman Mdallal

Abstract

The undisputed trait of modern technology is its vulnerability to obsolescence. This has been true in the past decades, where such technologies can be freshly introduced one day, while the next is surpassed by a more popular and more advanced version. With this in mind, this chapter compiles the latest ideas and available technologies in sustainable additive manufacturing. The authors spared no effort in ensuring that the reader is exposed to cutting-edge technologies during the time of writing this book. The objective, as with other parts of this book, is to provoke the imagination of the reader into trying out these technologies and to be inspired by them to have their own experience applied to their projects. Moreover, the novelty is assessed in the contribution to various newly charted fields for additive manufacturing, such as medical and culinary applications.

A. H. Alami (✉) · M. Ayoub · A. Yasin
Sustainable and Renewable Energy Engineering Department, University of Sharjah, 27272 Sharjah, United Arab Emirates
e-mail: aalalami@sharjah.ac.ae

M. Ayoub
e-mail: mohamad.ayoub@sharjah.ac.ae

A. Yasin
e-mail: ahmad.yasin@sharjah.ac.ae

H. Aljaghoub
Management Science and Engineering Department, Khalifa University, 127788 Abu Dhabi, UAE
e-mail: 100065161@ku.ac.ae

A. Alashkar
Ph.D. Program in Materials Science and Engineering, American University of Sharjah, 26666 Sharjah, United Arab Emirates
e-mail: b00028197@aus.edu

A. Mdallal
Sustainable Energy and Power Systems Research Centre, RISE, University of Sharjah, 27272 Sharjah, United Arab Emirates
e-mail: Ayman.Mdallal@sharjah.ac.ae

1 Powder Bed Fusion Technology

An established additive manufacturing (AM) technique that has evolved through extensive research and industrial advancements. These technological advancements have led to the widespread adoption of PBF technologies in commercial settings. Most specifically, PBF technologies are employed to manufacture high-value items that are not easily fabricated using traditional manufacturing methods. Moreover, they are utilized to manufacture low-volume, intricate, and complex designs. In most cases, these designs are demanded by the medical and aerospace fields. Despite the considerable developments achieved in PBF technologies, extensive research is lacking in areas related to the refining processes, specific materials used in the PBF process, and toolpath optimization for mitigating thermal issues (Cai & Zhou, 2022). Similar to other additive manufacturing technologies, PBF possesses a wide range of advantages. Such as the reduced costs associated with assembly processes and product customization, and the fabrication of products without the usage of structural support, enabling the production of extremely complex designs (Goodridge & Ziegelmeier, 2017). Even though production using PBF is longer than other additive manufacturing technologies, it can produce products with superior surface finish using minimal post-processing. Furthermore, PBF can manufacture multiple parts simultaneously to utilize the space of the building chamber fully (Vayre et al., 2012).

PBF employs a laser or thermal print head as a heat source to fuse a powdered material into 3D products. This process (shown in Fig. 1) continues by applying the heat source to particles in a powder bed, building up the layers (Goodridge & Ziegelmeier, 2017).

Typically, the powder bed is placed in an inert or partially vacuumed environment to protect the melted metal. A laser or electron beam then systematically scans each layer of the spread-out powder, selectively melting material according to the cross-section defined in the digital model of the

A. H. Alami (ed.), *Sustainable Additive Technologies and Innovative Manufacturing Techniques*, Advances in Science, Technology & Innovation, https://doi.org/10.1007/978-3-031-73657-5_6

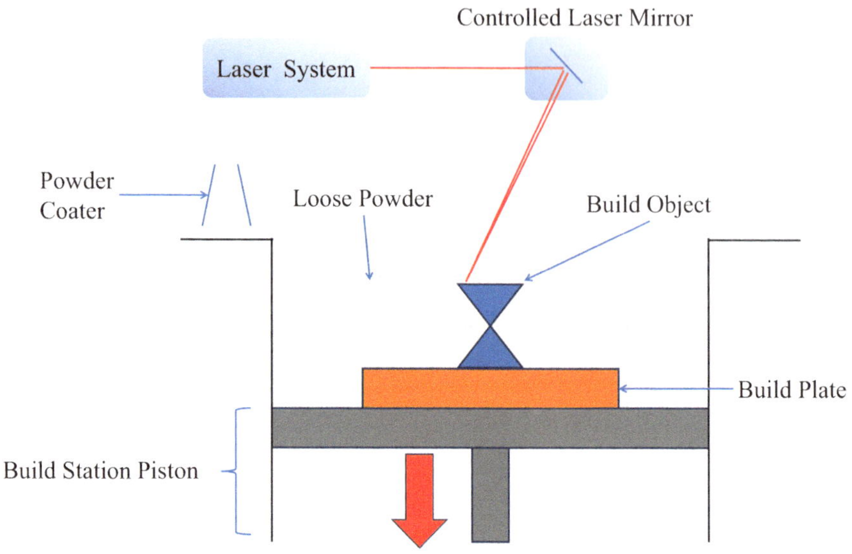

Fig. 1 Schematic of the PBF process

part. Once a layer is scanned, the building chamber's piston lowers while the powder chamber's piston rises by a predetermined thickness (Bhavar et al., 2014). The standard height of each layer in the PBF process ranges from 25 to 60 μm (O'Dowd & Todd, 2022). After this stage, the powder is deposited in the build chamber through a coating mechanism (mainly a roller), prior to being scanned by the energy source. This recurring process is carried out for each layer until the entire product is manufactured. This process results in a powder cake which conceals the shape of the final product. Hence, the last step involves removing the excess powder to display the geometry of the formed product (Bhavar et al., 2014). PBF technology naturally necessitates the use of an additional supporting structure to hold overhanding features in the component design. The material of the support structure must be identical to that of the manufactured part. The support structure is essential to prevent melted material from collapsing when there are overhanding surfaces. Support structures can be adjusted based on the part during the pre-processing stage. Meanwhile, the support is removed through mechanical treatment during the post-processing stage (Vayre et al., 2012). After this stage, the part can go through multiple treatments, such as polishing, shot peening, heat treatment, and machining (Bhavar et al., 2014).

In general, the PBF technique primarily utilizes a laser or an electron beam as an energy source. For instance, Electron Beam Melting (EBM) is the most common electron beam technology. While, Selective Laser Sintering (SLS), Selective Laser Melting (SLM), and Direct Metal Laser Sintering (DMLS), are considered the most utilized laser-powered sources (Dev Singh et al., 2021). The PBF processes are widely sought due to their capability to produce intricate parts with precise geometries in comparison to extrusion-based additive manufacturing technologies. Even though these processes generate inherently significant amounts of non-fused powder residue, these residues can be recycled for future usage (O'Dowd & Todd, 2022). Despite the substantial similarities in the working mechanism of these processes, SLS and SLM differ in the method of applying energy to solidify the metal particles. For instance, in the SLS process, the metal particles are heated until they are sintered, whereas the SLM process works on melting the metal particles. In a similar way, EBM uses an electron beam for heating the metal particles.

1.1 Selective Laser Sintering (SLS)

Carl Deckard introduced the SLS technique in 1986 and it was commercialized later in 1992 (Bhavar et al., 2014). SLS is an extremely adaptable technique capable of producing intricate geometries of high quality directly from CAD models. This method offers precision within a range of ±0.3% with a minimum layer thickness of 0.08 mm. Polyamide (PA), glass-filled polyamide (PA-GF), and alumide are the commonly employed materials in this process (Singh et al., 2020). SLS employs a laser beam to solidify and illuminate the metal powder one layer at a time, enabling the creation of intricate 3D parts. Initially, the powder material must be preheated to a specific operating temperature. This temperature should not exceed the material's softening temperature to prevent powder particles from adhering to each other. Selective solidification of the powder material is achieved through additional laser heating up to reach the 'sintering temperature'. Sintering happens as the powder temperature decreases and surface tensions increase, resulting in an interfacial kinetic process among the grains.

Generally, the SLS process follows three main steps to manufacture any part. Initially, a recoating roller applies a slim layer of metal powder on the surface of the building chamber. Subsequently, the laser beam is directed to selectively scan and sinter the first layer of powder material according to the digital model, creating the initial sintered layer. After completing the scan of each layer, the part's piston lowers by one layer, and simultaneously, the coating roller applies the next layer of powder onto the previously sintered layer. This process repeats until the desired part is fully manufactured (Han et al., 2022).

1.2 Selective Laser Melting (SLM)

In 1999, Fockele and Schwarze introduced the SLM technique. Which is perceived as an advanced version of the SLS process. The main difference lies in the fact that the powder particles are completely melted through the use of one or more laser beams. In comparison to SLS, the SLM technology has been developed to incorporate two sets of leasers, and each set contains two lasers, resulting in a total of four laser beams scanning the powder particles at the same time (Bhavar et al., 2014). Moreover, SLM operates at a faster pace, but needs inert gases for operating the laser beam. Hence, SLM is mostly utilized for manufacturing metal parts as it melts rather than sinters the powder particles (Singh et al., 2020).

Similar to all AM technologies, the printing begins with a CAD source file of the desired part. Subsequently, specialized software divides the model into cross-sections, which are then integrated into the manufacturing process (Yadroitsev et al., 2007). The critical stage in this process is using a laser mounted on top of lenses focused on the powder bed to solidify each layer. After melting the first layer, the build chamber descents while the recoating mechanism applies a new layer that needs to be solidified, until the until the entire part is completed. The thickness of the powder layer must be maintained at least at 0.02 mm. Recently, the SLM process has been employed to process diverse metallic materials, like aluminum, copper, and tungsten, using advanced high-power lasers utilizing fiber optics (Singh et al., 2020). The SLM technology possesses numerous advancements over conventional AM technologies, such as reduced production time for manufacturing complex parts and finer microstructure at high cooling rates (Wohlers, 2005). Furthermore, SLM produces parts that exhibit outstanding mechanical properties (Meiners et al., 1997; Over et al., 2001). For this reason, the SLM process is used to manufacture parts for the aerospace, medical, and automotive sectors (Lu et al., 2020).

1.3 Direct Metal Laser Sintering (DMLS)

Shortly after the introduction of the SLS method, the DMLS technique was developed. Similar to the SLM technology, DMSL is a laser-based AM technology that solely uses metal powders to manufacture different products. DMLS employs a single-layer process controlled by a computer-based laser beam to form its final products. In essence, the DMLS machine creates parts within a movable chamber by adding thin layers of powder, each around 0.1 mm thick. A powerful laser then precisely melts the metal powder in the powder bed based on a digital model. This layer-by-layer process continues until the product is fully formed as per the specified design. DMLS allows for control over the porosity of each layer, though complete elimination of porosity is not possible (Singh et al., 2020). This technology is used to create functionally graded materials, small-batch components, and functional prototypes. It also enables efficient production of aerospace parts with reduced raw material usage and lower costs (Dev Singh et al., 2021).

1.4 Electron Beam Melting (EBM)

EBM is an AM technology that emerged in 1997. Among the latest technologies in this field, EBM utilizes an electron gun to manufacture 3D parts from powder material (mainly metals). The key constraints of the EBM process include the layer thickness, diameter of focus, powder heating temperature, electron beam power, and electron beam current. The powder layer thickness is typically set between 50 and 200 μm (Gong et al., 2014). Furthermore, various metals can be used, including titanium alloys, titanium aluminide, stainless steel, Inconel, tool steel, aluminum alloys, copper, beryllium, and cobalt chrome (Udroiu, 2012). The most commonly utilized material in EBM is titanium, and by optimizing the layer thickness the production time, and potential costs can be reduced by approximately 35% (Gokuldoss et al., 2017).

EBM is a powder fusion process that uses an electron beam as the primary heat source to melt metal powders. The electron beam is emitted from a heated tungsten filament at high velocity and is controlled by two magnetic fields: the focusing coil and the deflecting coil. The focusing coil shapes the electron beam into a narrow diameter of 0.1 mm, while the deflecting coil directs the beam to specific points to scan the powder layer. Alloy powders are initially deposited in the build chamber using powder hoppers, and the chamber maintains a high vacuum atmosphere at very high temperatures. The focused electron beam melts the powder particles into layers by converting the electrons' kinetic energy into thermal energy, scanning each layer to create a 2-dimensional slice of the part (Romedenne et al., 2020). Each layer of the powder bed undergoes two scanning stages: preheating and melting. In the preheating stage, a high-current beam with a rapid scanning speed preheats the powder layer using multiple passes. During the melting stage, a low-current beam with a slower scanning speed melts the powder. After each layer is scanned, the build chamber lowers, and a new layer of alloy powder is deposited on top of the previously solidified layer. This process is repeated until the part is fully formed. Once the part is completed and removed, any unmelted metal powder is cleaned and recycled for future use (Romedenne et al., 2020).

2 Direct Energy Deposition Technology

Direct Energy Deposition (DED) technology is a laser-power-based process that was originally introduced in 1997 under the name Laser Engineered Net Shaping (LENS) (Svetlizky et al., 2022). After some time, organizations started offering DED machines globally. Meanwhile, other organizations started selling hybrid additive manufacturing machines combining DED and Wire and Arc Additive Manufacturing (WAAM) machines using filament feeding and have expanded their capabilities within this market segment. DED is perceived as a flexible metal additive manufacturing technology used for the direct production of a metal product based on a 3D digital model. DED technology is employed to fabricate any product on both even and uneven substrates (Ahn, 2021). The most commonly utilized DED processes use powder as feedstock and a laser beam as the energy source and is known as a Laser Additive Manufacturing DED process. (Svetlizky et al., 2021). Whereas if a metal wire is utilized as a feedstock in the DED process, then an electric arc, an electron beam, a plasma arc, or a laser can be used to produce thermal energy (Ahn, 2021). Figure 2 shows the DED process. The electron beam melts the feedstock at a specific point, forming a molten pool usually protected by an inert atmosphere. Similar to other AM technologies, DED employs a CAD software model to 3D print the desired part (DebRoy et al., 2018).

Although utilizing the wire form of the metal rather than the powder is less expensive and can be easily and safely stored, it requires higher laser power to melt and thus increases the cost of the overall system. In case powder is used as the main feedstock, the power deposition efficiency can be greatly improved by utilizing a coaxial feed nozzle. In general, the feedstock material can be made of alloy steels comprising titanium, nickel, aluminum, stainless steels, or tool steels. These are the conventional materials used as feedstock for DED machines. However, emerging materials are now processed using the DED machine, such as shape memory alloys, cobalt-based alloys, high entropy alloys, inter-metallics, ceramics, functionally graded materials, and composites (Svetlizky et al., 2022).

Although the core principles of DED are similar to those of welding, DED still offers distinct technical and scientific setbacks (Karayel & Bozkurt, 2020). Therefore, leveraging knowledge from welding metallurgy can enhance DED processes, ensuring a more controlled microstructure (Tan et al., 2020). Thus, integrating various welding techniques with additive manufacturing processes expands the potential of these processes for large-scale applications. However, even as stand-alone technologies, DED machines still offer numerous advantages to the additive manufacturing industry. Many DED machines currently operate as open-loop systems. While their fundamental operational concepts are consistent, each supplier incorporates distinct features into their machines. Furthermore, most vendors are currently focused on offering six-axis or free-axis freedom deposition heads, as they enable printing to be carried out on any surface and add new characteristics to finished products. On top of these advantages, these machines can print large free-standing structures (Svetlizky et al., 2022).

Even though PBF generally offers superior dimensional precision and produces end products with less surface

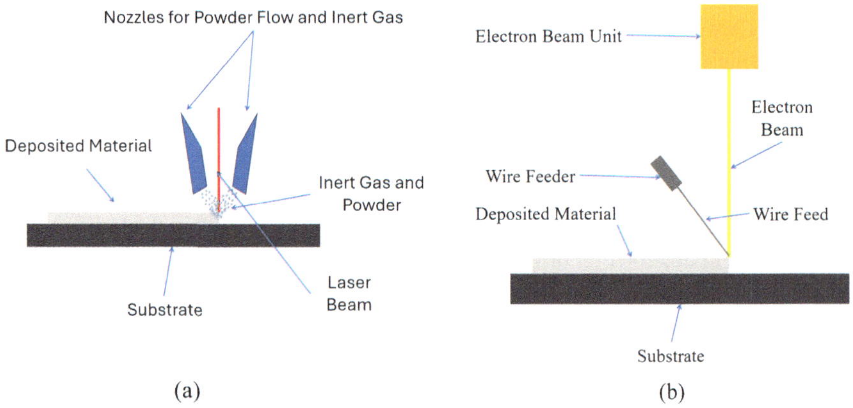

Fig. 2 Direct energy deposition process for **a** powder and **b** wire feedstock

roughness (DebRoy et al., 2018; Svetlizky et al., 2021), DED boasts several distinct benefits over PBF processes. These benefits are summarized in the following points:

- A higher deposition rate, of around 2.5 kg/h, in comparison to 0.25 kg/h for PBF processes.
- Inherently supports multi-material deposition, including in situ alloying, and enables the creation of functionally graded parts with customized properties at specific locations (Heer & Bandyopadhyay, 2018; Noecker & Dupont, 2002).
- Employment in a multitude of sectors such as: cladding, coating, and repairing scraped parts (Saboori et al., 2019; Wilson et al., 2014).
- DED accommodates the fabrication of large parts with volumes exceeding 1000 mm^3.
- DED machines with more than 5 axes can enable layer deposition in all directions.

3 Materials for Sustainable Additive Manufacturing

3.1 Polymeric Materials from Organic and Recycled Plastic

Turning household-produced waste plastic into filaments has had a spark of interest due to the ease of personal filament production and facilitating plastic recycling, thus reducing the need for new plastic production. A company called FilaBot has designed a tabletop device that takes household plastic waste such as bottles and bags or even obsolete broken 3D parts, grinds them into little pallets, and feeds them into an extruder, that subsequently heats and melts the plastic and forces it into a nozzle to create a continuous strand of filament. This is finally cooled and wound onto a spool. Some plastic types may require a further drying step to remove any moisture residue from the filament before it can be used for 3D printing applications. The produced filament can either be 3 or 1.75 mm in size, where a single plastic bottle of milk can produce up to 2.5 m of filament (*Filabot—Recycling for 3D Printing*, n.d.; *Filabot Turns Your Plastic Junk Into Material for 3-D Printers | WIRED*, n.d.). The overall process is highlighted in Fig. 3. High-density polyethylene (HDPE), Low-density polyethylene (LDPE), Acrylonitrile butadiene styrene (ABS), and Nylon are different types of "safe" plastics that are suitable feeds for such an application.

The efficiency of producing filaments from plastic waste relies on numerous aspects such as the quality of the grinder and extruder that are used in the system, but most importantly, the purity of the plastic waste that is used. For instance, if the plastic waste is contaminated, it may cause issues to arise during the extrusion process, such as clogging, resulting in a lower quality filament product.

From an environmental point of view, recycling plastic into 3D printing filaments has an overall positive impact, which is justified by the reduced amount of plastic that ends up in landfills. Moreover, recycled plastic attenuates the effects of carbon dioxide emissions from newly produced plastic, due to the lower energy requirements to recycle old plastic in comparison to producing new plastic. However, it is worth noting that a positive environmental impact is not imminent in all cases. If the recycled plastic contains harmful chemicals, such as polyvinyl chloride (PVC), it may produce harmful fumes and emissions that can be a health risk to the personnel and adverse effect on the environment. Finally, if the produced filament is not biodegradable and ends up in a landfill at the end of its lifetime, it will take hundreds, if not thousands of years, to decompose.

Plant-based 3D filaments are filaments that are produced from soy, starch, algae, and wood. Plant-based 3D filaments are considered lucrative alternatives to petroleum-based ones due to many advantageous characteristics such as biodegradability, which causes little to no harm to the environment by the end of the filament's lifecycle, reduced carbon footprint as the sources that make up plant-based 3D

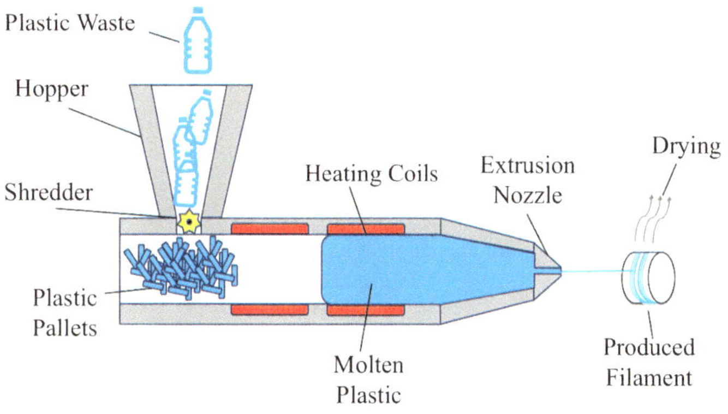

Fig. 3 Plastic waste to filament production

filaments naturally absorb carbon dioxide (CO_2), and the natural diverse finishes of the produced filaments. Plant-based filaments fall under the category of Polylactic acid (PLA). Generally, PLA sources are dictated geographically, depending on the abundant plant sources that are available in specific regions, for instance, in the USA or Canada the main PLA source would be corn (starch), in Asia roots and chips, and in the rest of the world sugar cane, wheat and potatoes (*SelfEco Breaks Down PLA: Making 3D Printing Eco-Friendly—3D Printing Industry*, n.d.). Moreover, given that around 17% of global food production is wasted (Nations, n.d.), it can be a rich source for filament production.

Producing PLA is done through a process that starts with waste food (raw source) and ends with usable 3D filaments. The initial step of PLA production is lactic acid fermentation. Various fungi are added to food waste, where enzymes are secreted in situ, which essentially forms the lactic acid feedstock. Bacteria is added to the food waste mixture, where it consumes the nutrients available to produce lactic acid. Next comes lactic acid recovery, where cell biomass and other impurities are removed via centrifuge and active carbon absorption, respectively. Subsequently, the recovered lactic acid is turned into a polymer via ring-opening polymerization. Then, a metal catalyst is introduced for lactide synthesis, followed by a further purification step. The formed polymer can either be turned into pallets, or into filaments which is achieved via melt spinning. The overall process is shown in Fig. 4.

Soy-based filaments are a type of PLA 3d printing filaments that consist of approximately 20–25% soy-based materials (*Filasoy: Green, Soy Based 3D Printer Filament Created By Purdue Students—3DPrint.Com|The Voice of 3D Printing/Additive Manufacturing*, n.d.). It is feasible to manufacture such filaments in geographical areas where soy is a large part of the local economy, such as Indiana, USA, where soy-based filaments were first produced (FilaSoy). Soy-based filaments have a lot of advantages over conventional PLA materials including reduced brittleness, and retaining the quality of the printed objects, all with a lower filament melting temperature. Moreover, soy-based filaments have anti-microbial properties which makes them a better choice for prints that may come into contact with bacteria. A soy-based biodegradable filament for a fused filament fabrication (FFF) application was developed (Dey et al., 2023), where a soy-bean processing product, soy hulls, was blended in a PLA matrix. The lightweight low-cost soy hull resulted in a reduced filament density and cost. Moreover, they incorporated a poly 2-ethyl-2-oxazoline (PEOX) additive to enhance the soy hull-PLA bonding. A bioink developed by Temple University consisting of soy protein isolate (SPI—90% protein) was extruded to construct complex edible geometries that are used for food production, where ingredients can be extruded layer by layer to obtain various textures, flavors and whole meals that can be compared to regular food, and 3D printed scaffolds for tissue regeneration (Lelkes & Schwab, n.d.).

Algae are fast-growing marine organisms that are capable of producing vast amounts of biomass in a notably short period of time. Algae-based 3D printing which are extrusion, inkjet, and laser-assisted bioprinting. Compared to conventional PLA filaments, algae-based filaments (i.e., Algix) are

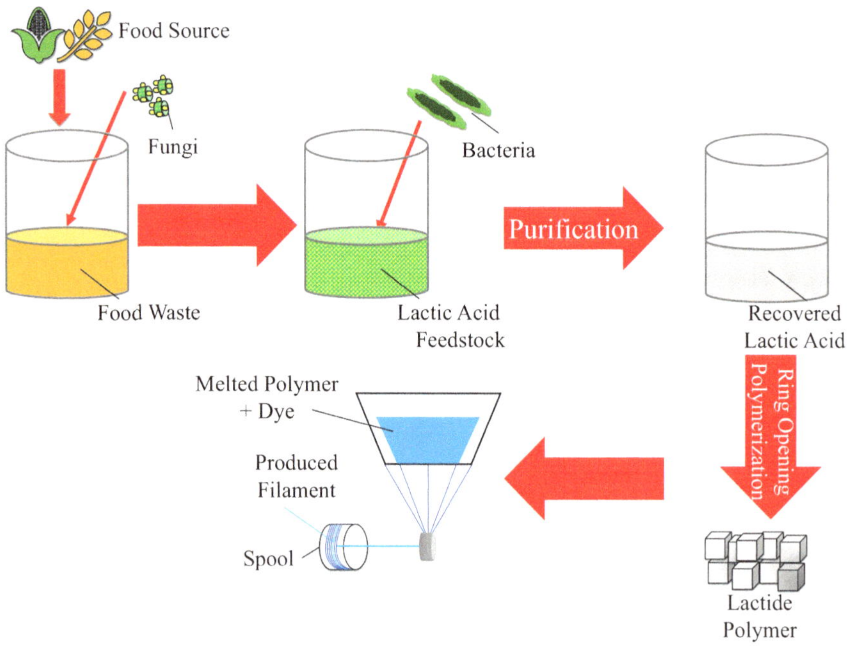

Fig. 4 Food to filament (PLA) production

favored due to their sustainability, given the rate of biomass produced in comparison to the respective greenhouse gasses emitted. Additionally, the biodegradability of algae-based filaments, makes them environmentally friendly as they break down within the environment over time. Furthermore, the versatility of printing algae-based materials allows for their utilization in many various fields, especially in the medical sector. There are multiple algae-based materials that are suited for producing 3D printing filaments for medical purposes, such as alginate, agarose, and carrageen (Liu et al., 2019). Algae-based 3D printing facilitates the printing of 3D geometries that are superior to their 2D counterparts in applications such as research conducted on natural cells-biomaterials interactions, understanding defects in tissues, drug testing, bioprinted vascular tissues, bone printing, and cartilage printing (Axpe & Oyen, 2016).

3.2 Medical Applications

Sustainable additive manufacturing has been on the rise due to newly found delicate applications and their requirements, such as those medical based. Through sustainable materials, the healthcare sector can be set to achieve unprecedented levels of customization to meet the demands mandated by prosthetics and implants. There are various feedstock sources that can be used for sustainable development and manufacturing of medical-related parts, and as mentioned earlier, these sources harbor a wide arsenal of materials ranging from plant-based to those recycled plastics. Moreover, biocompatibility is paramount when it comes to filaments used for medical additive manufacturing in order to meet health and safety regulations, as well as to adhere to profound ecological restrictions. Straying from a material-oriented perspective, specific medical applications mandate the use of certain additive manufacturing techniques that best suit their case. From the myriad of techniques mentioned earlier in this chapter, photo-polymerization is mostly used for bone and dental models/implant guides, as well as hearing aids, material jetting is used for dental casts, binder jetting is used for anatomy color coding, extrusion is used to manufacture medical devices and rapid exoskeleton prototyping, power bed fusion is used to develop lattice-requiring models (i.e., implants and fixations), and direct energy deposition is currently constrained to repairing existing parts (Aimar et al., 2019). The pharmaceutical aspect of the healthcare industry has also benefitted from advances made in sustainable additive manufacturing, with a prosperous timeline of development which includes 3D printed placebo tablets, 3D printed drug loaded printed devices, bilayer tablets, as well asl Spritam® (the first 3D printed drug), all which utilize the aforementioned additive manufacturing techniques. In retrospect, however, all of these advances would have an elevated impact on the medical scene if the utilized materials had a "green" or a sustainable aspect to them, which is why green materials for medical applications have boomed in terms of research quantity and quality and continue to head towards more sustainable solutions (Jamróz et al., 2018). A filament comprising a polycaprolactone (known for its green nature) matrix housing zinc, copper, and silver, bearing antimicrobial wound dressing, was developed and printed through the shapes of ears and noses, which would ensure a continuous release of these transition metals on the interface with a wound to maintain the antibacterial exposure for the maximum effective duration (Muwaffak et al., 2017). Moreover, poly ether-ether-ketone (PEEK) has caught attention due to its superior lignin biocompatibility with other types of polymers, which suit biomedical applications and related materials, such as cellular calcium hydroxyapatite. This material has been successfully utilized to develop an intervertebral cage, which due to its luminescence, can help with visualizing the bone growth within through x-ray. Given its polymeric characteristics and stress–strain behavior, this material makes for a suitable 3D printing feedstock (Oladapo et al., 2021).

3.3 Food Applications

Animal agriculture, a primary source of feedstock for the food industry, is responsible for 65% of global nitrous oxide emissions, which have 296 times the impact of carbon dioxide (*The Animal Agriculture Industry and Zero Waste*, n.d.). Emerging 3D printing technologies have proven their ability to produce food with complex and delicate structures, combined with reduced cost and energy requirements. For example, meat substitutes can be produced using materials derived from vitro cell cultures, meat byproducts, insects, or even plants. The process of meat 3D printing involves the extrusion of fibrous meat materials using different extruder types such as screw conveyer or syringe system, which has great potential in reaching the desired composition and 3D structure with proper temperature control (Ramachandraiah, 2021). In terms of materials, there are various aspects to consider while utilizing 3D printing for food applications, which include the addition of edible binder materials, flavor additions, primary material feedstock, as well as the physiochemical properties of the materials involved. 3D food printing is not restricted to meat-based products, where this technology can also be utilized in printing fruit seeds, that would be normally wasted, into a consumer-friendly product. Jackfruit seeds have demonstrated high printability potential with little to no other additives, which maximizes their benefits as a "waste" feedstock for food 3D printing. These seeds are boiled for around 40 minutes until they are tinder. Then, their shells are peeled away, cooled, and

Fig. 5 The 3D printed jackfruit seed ink—foodini printer (4-mm nozzle)—adapted from (Hooi Chuan Wong et al., 2022) (with permission no: 5814811258253)

ground to paste, where water is added in a final step to form the food ink. 3D food printers, such as Foodini, are utilized to manufacture 3D printed parts, as shown in Fig. 5, from the obtained Jackfruit seed ink (Hooi Chuan Wong et al., 2022).

Food containers and holders are also essential when dealing with the supply chain of food products, and luckily there are advancements being made to make these factors more sustainable and 3D printable. A composite made from PLA (70%) and calcium carbonate ($CaCO_3$) (30%) has been used to produce a 3D printing-compatible filament for disposable food contact applications. It can be seen from Fig. 6, that with a 20% fill, the print patterns are much more obvious in comparison to the 80% samples. However, the 80%-filled samples had bubbles forming on their surface, which was absent in the case of the 20%-filled samples. Moreover, utilizing a triangular pattern also results in a different surface finish in comparison to the rectilinear pattern (Faba et al., 2024).

3.4 Construction Applications

Utilizing AM in construction offers the opportunity to delve into a diverse range of feed materials tailored to various parts of a building structure. One of the most intriguing aspects of 3D-printed buildings is the construction of concrete walls. This structure accommodates the use of various concrete mixtures. The versatility in feed material enables the incorporation of sustainable and recycled materials as a filament, promoting flexibility to explore more environmentally friendly construction materials. A recent study done on using ordinary port-land cement, recycled aggregate-based cement, and geopolymers as filaments for 3D concrete printing, found that not only does 3D cement printing have a positive environmental impact due to being able to use more environmentally friendly feed and utilizing all of the feed material with little to no excess matter, but it also promises a cost-effective solution due to less material being wasted when compared to the conventional construction method (Alami et al., 2023).

One of the most commonly used materials in the interior design industry is wood which generates a significant amount of waste during its processing. To produce an extrusion-based filament, wood waste such as slabs, shavings, and sawdust are milled into baking flour consistency and then mixed in with a binding polymer. The mixture is then extruded as a filament to be used in a variety of applications such as producing furniture elements. This process promotes a circular economy by aiding in limiting the volume of byproducts ending up in landfills or getting burned (Pringle et al., 2018; Romani et al., 2021).

Gypsum plaster, commonly used for wall and ceiling decoration in interior design, also holds potential as an extrudable material for 3D printing using a motor-driven syringe extruder. While powder bed binder jetting technology for shaping gypsum is already commercialized, extrusion-based 3D printing presents challenges. Factors such as setting time, hydration heat, buildability, and viscoelasticity are critical for evaluating the final product. For successful extrusion printing, the standard gypsum-water

Fig. 6 Resultant prints using PLA-$CaCO_3$ Composite for food contacts—adapted from (Faba et al., 2024) (Open Access)

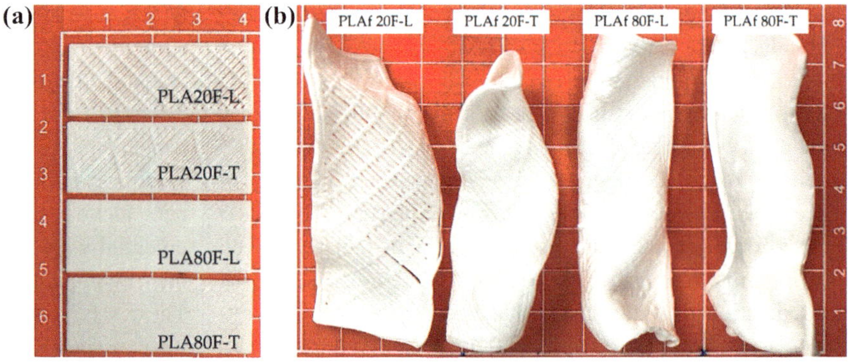

mixture requires additives like retarders and heat-induced accelerators to maintain shape and facilitate proper extrusion. Economically, gypsum is an appealing choice for filament material due to its abundant availability from natural sources and industrial by-products, making it a reasonably sustainable option (Huang et al., 2021).

References

Ahn, D.-G. (2021). Directed energy deposition (DED) process: State of the art. *International Journal of Precision Engineering and Manufacturing-Green Technology, 8*, 703–742.

Aimar, A., Palermo, A., & Innocenti, B. (2019). The role of 3D printing in medical applications: A state of the art. *Journal of Healthcare Engineering, 2019*, 1–10. https://doi.org/10.1155/2019/5340616

Alami, A. H., Olabi, A. G., Ayoub, M., Aljaghoub, H., Alasad, S., & Abdelkareem, M. A. (2023). 3D concrete printing: Recent progress, applications, challenges, and role in achieving sustainable development goals. *Buildings, 13*(4), 924. https://doi.org/10.3390/BUILDINGS13040924

Axpe, E., & Oyen, M. (2016). Applications of alginate-based bioinks in 3D bioprinting. *International Journal of Molecular Sciences, 17* (12), 1976. https://doi.org/10.3390/ijms17121976

Bhavar, V., Kattire, P., Patil, V., Khot, S., Gujar, K., & Singh, R. (2014). *A review on powder bed fusion technology of metal additive manufacturing*.

Cai, C., & Zhou, K. (2022). Metal additive manufacturing. In: *Digital manufacturing* (pp. 247–298).

DebRoy, T., Wei, H. L., Zuback, J. S., Mukherjee, T., Elmer, J. W., Milewski, J. O., Beese, A. M., Wilson-Heid, A., De, A., & Zhang, W. (2018). Additive manufacturing of metallic components—Process, structure and properties. In: *Progress in materials science* (vol. *92*, pp. 112–224). https://doi.org/10.1016/j.pmatsci.2017.10.001

Dev Singh, D., Mahender, T., & Raji Reddy, A. (2021). Powder bed fusion process: A brief review. *Materials Today: Proceedings, 46*, 350–355. https://doi.org/10.1016/j.matpr.2020.08.415

Dey, A., Rahman, M. M., Yodo, N., & Grewell, D. (2023). Development of biocomposite filament for fused filament fabrication from soy hulls and soy protein isolate. *Materials Today Communications, 34*, 105316. https://doi.org/10.1016/j.mtcomm.2023.105316

Faba, S., Agüero, Á., Arrieta, M. P., Martínez, S., Romero, J., Torres, A., & Galotto, M. J. (2024). Foaming of 3D-printed PLA/$CaCO_3$ composites by supercritical CO_2 process for sustainable food contact materials. *Polymers, 16*(6), 798. https://doi.org/10.3390/polym16060798

Filabot—Recycling for 3D Printing. (n.d.). Retrieved June 23, 2024, from https://www.azom.com/article.aspx?ArticleID=8114

Filabot Turns Your Plastic Junk Into Material for 3-D Printers| WIRED. (n.d.). Retrieved June 23, 2024, from https://www.wired.com/2013/01/filabot-plastic-recycler/

Filasoy: Green, Soy Based 3D Printer Filament Created By Purdue Students—3DPrint.com|The Voice of 3D Printing/Additive Manufacturing. (n.d.). Retrieved June 23, 2024, from https://3dprint.com/1835/filasoy-green-soy-based-3d-printer-filament-created-by-purdue-students/

Gokuldoss, P. K., Kolla, S., & Eckert, J. (2017). Additive manufacturing processes: Selective laser melting, electron beam melting and binder jetting-selection guidelines. *Materials*.

Gong, X., Anderson, T., & Chou, K. (2014). Int. Review on powder-based electron beam additive manufacturing technology. *Manufacturing Review*. https://doi.org/10.1051/mfreview/2014001

Goodridge, R., & Ziegelmeier, S. (2017). Powder bed fusion of polymers. In *Laser additive manufacturing* (pp. 181–204). Elsevier.

Han, W., Kong, L., & Xu, M. (2022). Advances in selective laser sintering of polymers. *International Journal of Extreme Manufacturing, 4*(4), 42002.

Heer, B., & Bandyopadhyay, A. (2018). Compositionally graded magnetic-nonmagnetic bimetallic structure using laser engineered net shaping. *Materials Letters, 216*, 16–19. https://doi.org/10.1016/j.matlet.2017.12.129

Hooi Chuan Wong, G., Pant, A., Zhang, Y., Kai Chua, C., Hashimoto, M., Huei Leo, C., & Tan, U. -X. (2022). 3D food printing—Sustainability through food waste upcycling. *Materials Today: Proceedings, 70*, 627–630. https://doi.org/10.1016/j.matpr.2022.08.565

Huang, J., Duan, B., Cai, P., Manuka, M., Hu, H., Hong, Z., Cao, R., Jian, S., & Ma, B. (2021). On-demand setting of extrusion-based 3D printing gypsum using a heat-induced accelerator. *Construction and Building Materials, 304*, 124624. https://doi.org/10.1016/j.conbuildmat.2021.124624

Jamróz, W., Szafraniec, J., Kurek, M., & Jachowicz, R. (2018). 3D printing in pharmaceutical and medical applications—Recent achievements and challenges. *Pharmaceutical Research, 35*(9), 176. https://doi.org/10.1007/s11095-018-2454-x

Karayel, E., & Bozkurt, Y. (2020). Additive manufacturing method and different welding applications. *Journal of Materials Research and Technology, 9*(5), 11424–11438.

Lelkes, P., & Schwab, K. (n.d.). *Sustainable 3D Printing with Soy-derived Bioink*.

Liu, J., Sun, L., Xu, W., Wang, Q., Yu, S., & Sun, J. (2019). Current advances and future perspectives of 3D printing natural-derived biopolymers. *Carbohydrate Polymers, 207*, 297–316. https://doi.org/10.1016/j.carbpol.2018.11.077

Lu, Q. Y., Nguyen, N. V., Hum, A. J. W., Tran, T., & Wong, C. H. (2020). Identification and evaluation of defects in selective laser melted 316L stainless steel parts via in-situ monitoring and micro computed tomography. *Additive Manufacturing, 35*, 101287.

Meiners, W., Wissenbach, K., & Poprawe, R. (1997). Direct selective laser sintering of steel powder. *Proceedings of the LANE, 97*, 615–622.

Muwaffak, Z., Goyanes, A., Clark, V., Basit, A. W., Hilton, S. T., & Gaisford, S. (2017). Patient-specific 3D scanned and 3D printed antimicrobial polycaprolactone wound dressings. *International Journal of Pharmaceutics, 527*(1–2), 161–170. https://doi.org/10.1016/j.ijpharm.2017.04.077

Nations, U. (n.d.). *Food Loss and Waste Reduction|United Nations*. Retrieved June 23, 2024, from https://www.un.org/en/observances/end-food-waste-day

Noecker, F., & Dupont, J. (2002). Functionally graded copper—Steel using laser engineered net shaping process. In *ICALEO 2002 - 21st International Congress on Applications of Laser and Electro-Optics, Congress Proceedings*.

O'Dowd, N., & Todd, M. (2022). Impacts of metal additive manufacturing on smart city infrastructure. In *The rise of smart cities* (pp. 53–78). Elsevier.

Oladapo, B. I., Zahedi, S. A., Ismail, S. O., & Omigbodun, F. T. (2021). 3D printing of PEEK and its composite to increase biointerfaces as a biomedical material—A review. *Colloids and Surfaces B: Biointerfaces, 203*, 111726. https://doi.org/10.1016/j.colsurfb.2021.111726

Over, C., Meiners, W., Wissenbach, K., Lindemann, M., & Hammann, G. (2001). Selective laser melting: A new approach for the direct

manufacturing of metal parts and tools. *Proceedings of the International Conferences on LANE, 3*, 391–398.
Pringle, A. M., Rudnicki, M., & Pearce, J. M. (2018). Wood furniture waste–based recycled 3-D printing filament. *Forest Products Journal, 68*(1), 86–95. https://doi.org/10.13073/FPJ-D-17-00042
Ramachandraiah, K. (2021). Potential development of sustainable 3D-printed meat analogues: A review. *Sustainability, 13*(2), 938. https://doi.org/10.3390/su13020938
Romani, A., Rognoli, V., & Levi, M. (2021). Design, materials, and extrusion-based additive manufacturing in circular economy contexts: From waste to new products. *Sustainability, 13*(13), 7269. https://doi.org/10.3390/SU13137269
Romedenne, M., Pillai, R., Kirka, M., & Dryepondt, S. (2020). High temperature air oxidation behavior of Hastelloy X processed by Electron Beam Melting (EBM) and Selective Laser Melting (SLM). *Corrosion Science, 171*, 108647.
Saboori, A., Aversa, A., Marchese, G., Biamino, S., Lombardi, M., & Fino, P. (2019). Application of directed energy deposition-based additive manufacturing in repair. *Applied Sciences, 9*(16), 3316.
SelfEco Breaks Down PLA: Making 3D Printing Eco-friendly—3D Printing Industry. (n.d.). Retrieved June 23, 2024, from https://3dprintingindustry.com/news/mit-breaks-down-pla-making-3d-printing-eco-friendly-67710/
Singh, R., Gupta, A., Tripathi, O., Srivastava, S., Singh, B., Awasthi, A., Rajput, S. K., Sonia, P., Singhal, P., & Saxena, K. K. (2020). Powder bed fusion process in additive manufacturing: An overview. *Materials Today: Proceedings, 26*, 3058–3070. https://doi.org/10.1016/j.matpr.2020.02.635
Svetlizky, D., Das, M., Zheng, B., Vyatskikh, A. L., Bose, S., Bandyopadhyay, A., Schoenung, J. M., Lavernia, E. J., & Eliaz, N. (2021). Directed energy deposition (DED) additive manufacturing: Physical characteristics, defects, challenges and applications. *Materials Today, 49*, 271–295.
Svetlizky, D., Zheng, B., Vyatskikh, A., Das, M., Bose, S., Bandyopadhyay, A., Schoenung, J. M., Lavernia, E. J., & Eliaz, N. (2022). Laser-based directed energy deposition (DED-LB) of advanced materials. *Materials Science and Engineering: A, 840*, 142967. https://doi.org/10.1016/j.msea.2022.142967
Tan, J. H. K., Sing, S. L., & Yeong, W. Y. (2020). Microstructure modelling for metallic additive manufacturing: A review. *Virtual and Physical Prototyping, 15*(1), 87–105.
The animal agriculture industry and zero waste. (n.d.). Retrieved May 25, 2024, from https://www.colorado.edu/ecenter/2022/03/15/it-may-be-uncomfortable-we-need-talk-about-it-animal-agriculture-industry-and-zero-waste
Udroiu, R. (2012). Powder bed additive manufacturing systems and its applications. *Academic Journal of Manufacturing Engineering, 1583–7904*(10), 122–129.
URL link. https://www.3dprintlife.com/alga
Vayre, B., Vignat, F., & Villeneuve, F. (2012). Metallic additive manufacturing: State-of-the-art review and prospects. *Mechanics & Industry, 13*, 89–96. https://doi.org/10.1051/meca/2012003
Wilson, J. M., Piya, C., Shin, Y. C., Zhao, F., & Ramani, K. (2014). Remanufacturing of turbine blades by laser direct deposition with its energy and environmental impact analysis. *Journal of Cleaner Production, 80*, 170–178.
Wohlers, T. (2005). *Wohlers report: Rapid prototyping*. Wohlers Associates Inc.
Yadroitsev, I., Bertrand, Ph., & Smurov, I. (2007). Parametric analysis of the selective laser melting process. *Applied Surface Science, 253* (19), 8064–8069. https://doi.org/10.1016/j.apsusc.2007.02.088

Do-It-Yourself Project: Sand Casting and Lost PLA/Wax

Abdul Hai Alami, Adnan Alashkar, Mohamad Ayoub, Ayman Mdallal, Ahmad Yasin, Shamma Alabdulla, Fabrizia Devito, Siren Khuri, Haya Aljaghoub, and Fatima Alshamsi

Abstract

This chapter highlights the real motivation to embark on writing this book. The products that went from being a design concept into tangible metallic parts are presented as a practical example of the capabilities of additive manufacturing. The parts can be made of PLA as well as well as metal. With design and CAD knowledge, the skilled artisan is removed from the equation and anyone with a design idea and good CAD modeling knowledge can design, 3D print then cast customized functional parts. The trial and error that goes into such processes is what builds the know-how, and the more a person is involved in this process the deeper the experience and knowledge become.

A. H. Alami (✉) · M. Ayoub · A. Yasin · F. Alshamsi
Sustainable and Renewable Energy Engineering Department, University of Sharjah, 27272 Sharjah, United Arab Emirates
e-mail: aalalami@sharjah.ac.ae

M. Ayoub
e-mail: mohamad.ayoub@sharjah.ac.ae

A. Yasin
e-mail: ahmad.yasin@sharjah.ac.ae

F. Alshamsi
e-mail: u20200968@sharjah.ac.ae

A. Alashkar
PhD Program in Materials Science and Engineering, American University of Sharjah, 26666 Sharjah, United Arab Emirates
e-mail: b00028197@aus.edu

A. Mdallal · S. Khuri
Sustainable Energy and Power Systems Research Centre, RISE, University of Sharjah, 27272 Sharjah, United Arab Emirates
e-mail: Ayman.Mdallal@sharjah.ac.ae

S. Khuri
e-mail: SKhuri@sharjah.ac.ae

S. Alabdulla
Mechanical Engineering Department, American University of Sharjah, Sharjah, 26666, UAE
e-mail: g00070854@aus.edu

F. Devito
Department of Mechanical, Mathematics and Management (DMMM), Polytechnical University of Bari, Bari, Italy
e-mail: fabrizia.devito@poliba.it

H. Aljaghoub
Management Science and Engineering Department, Khalifa University, 127788 Abu Dhabi, UAE
e-mail: 100065161@ku.ac.ae

1 Introduction

Metal casting is a facile and scalable way to rapidly manufacture functional and durable parts. The idea is to pour molten metal into a mold that takes up the shape of the desirable product. There are two common techniques to create molds suitable for metal casting, which are sand and lost-wax casting.

Sand casting is a technique that utilizes a dummy part out of a cheap and non-durable material to create a mold. The dummy part is temporarily inserted into the sand and then removed, creating a cavity in the sand, which is encapsulated by steel frames, that perfectly takes its shape and details. There are multiple cautionary steps that are taken into account to ensure a successful metal pour, such as adding runners and manually creating channels within the two halves of the sand mold, which will be described in detail in the sections that follow.

Another technique that is commonly used is lost-wax casting. Similar to sand casting, a temporary part is created using wax carving, or via 3D printing in this case, which resembles the desired metal end part. The 3D-printed part is then secured on a removable base, and liquid plaster is poured on top. Once the base is removed, it exposes the bottom of the 3D-printed part allowing its subsequent firing and removal from the plaster mold, leaving behind a cavity in the plaster. The metal is finally poured into the hardened plaster where it takes the shape of the 3D printed part. This

A. H. Alami (ed.), *Sustainable Additive Technologies and Innovative Manufacturing Techniques*, Advances in Science, Technology & Innovation, https://doi.org/10.1007/978-3-031-73657-5_7

technique allows the casting of multiple items and shapes that can be attached to a single "tree" within the plaster mold.

In this chapter, sand and lost-wax casting are described in detail, highlighting their advantages and shortcomings, as well as preventative measures that can ensure a better casting process and overall experience.

2 Sand Casting: Principals, Equipment and Materials

The sand-casting process is one of the most ancient manufacturing processes that date back to as far as 1100 BC, where records have shown its use in ancient China. It basically involves making a cavity in the sand that has enough self-adhesion to allow the intricate details to be indented in the sand from a temporary pattern that is placed within the sand as the sand is compacted around its external surface. The process of producing the pattern, which was made of a material amenable to being formed into shape by material removal processes, required a skilled master to make the pattern take the required shape with as many details as possible. This pattern is then properly covered in sand, which was compacted around the pattern in such a way that would leave an impression in the sand after the pattern is removed. A molten material (mostly a nonferrous metal or alloy) flows in the cavity and is allowed to slowly solidify, taking the shape of the cavity left by the pattern. The pattern can then be reused, but the cavity will eventually collapse as the cast part is removed and the sand will have to be rehabilitated for subsequent casting processes. The general sequence of a sand casting process is explained in the following steps.

2.1 Pattern Preparation

The pattern is the sought-end product, but in order to produce the required end product via sand casting, the pattern has to be prepared under certain conditions that may not be intuitive. For example, if the cylindrical object shown in Fig. 1a is to be sand cast from aluminum, a cylindrical cavity must be produced within the sand. This entails burying a cylindrical object within highly compact sand and then removing it carefully, pouring molten aluminum within the cavity, allowing it to cool down, and then recovering the final part. The pattern of course can be more complicated than a simple cylindrical shape and thus should be fabricated by a skilled master from a malleable material. With the advent of CAD and additive manufacturing, the role of skill has evolved into a virtual product (file) that can be edited by several collaborators who can come from various departments or backgrounds to ensure the intended function of the part is achieved. As will be explained in detail later in the chapter, there are several provisions that must be added to the pattern to facilitate the removal process, such as dividing the pattern into two parts or adding alignment features such as locating pins, as shown in Fig. 1c. One interesting feature that should be considered when designing the pattern is compensating for metal shrinkage. If the sought metallic part has to be the same size as the pattern, the pattern has to be made larger in dimensions to compensate for shrinkage. The magnitude of shrinkage varies according to the metal used, the casting shape, and the pouring temperature. Adding 2–3% to the dimensions of the pattern is one way to compensate for shrinkage (Santhi, 2018), but also incorporating blind risers into the flow system design would help to physically compensate for the expected shrinkage as seen in Fig. 1b.

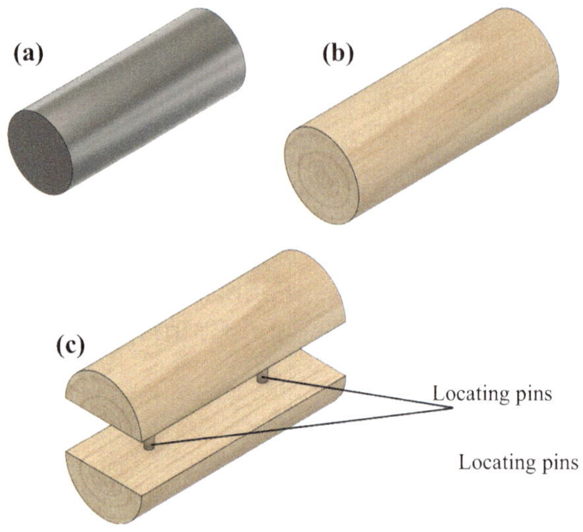

Fig. 1 **a** Sought part, **b** ideal pattern, **c** actual pattern that is cut half and has locating pins that facilitate alignment

2.2 Mould Preparation

Once the pattern is ready, a proper mould that is used to hold and compact the sand needs to be used. The mould (or flask) can be made from virtually any material, as it will not be under large loads itself, but it should support the compact clay that has, in turn, to contain a moving front of dense molten metal and retain surface details from the pattern that just left the cavity. A typical sand-casting mold is shown in Fig. 2.

There are several considerations when filling out the mould with the casting sand, but one of the main factors is the compactness of the sand around the pattern. For this to be accomplished, the pattern has to be cut into two parts, one housed in the drag (bottom mould part) and the other in the cope (top mould part). The drag is usually filled out first and this is done with it being upside down. The pattern with the smooth surface, i.e. top section of Fig. 1c, is placed on a flat

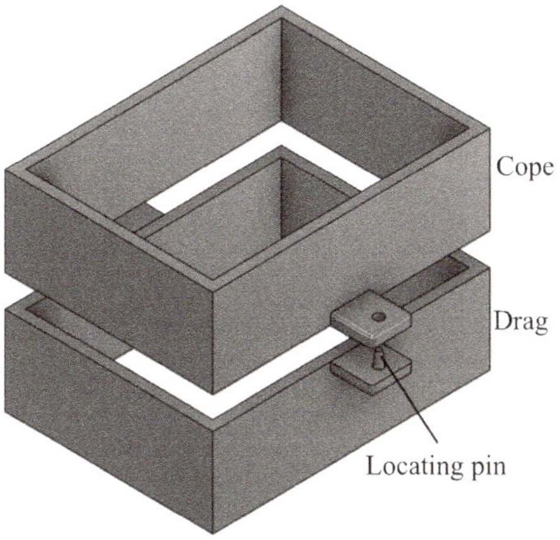

Fig. 2 Typical sand-casting mould

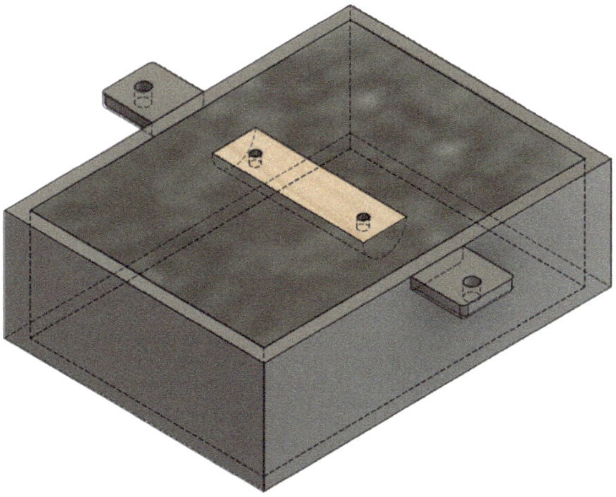

Fig. 3 The drag with half the pattern. Note the holes for the locating pins in the pattern (and also on the mould handles) to ensure perfect alignment

and straight surface facing down, and the clay is compacted around it, first by hand then with a compacting tool such as a plastic mallet. Talcum powder can be used to ensure that the pattern does not stick to the sand once it is removed from it. For now, the drag that is currently full of casting sand is flipped over, exposing the half pattern surrounded by clay, as shown in Fig. 3.

The next step involves fixing the cope over the drag with the locating pins, gluing the other half of the pattern on the one already surrounded by sand in the drag, and then adding sand to the cope cavity and compacting it. The pattern then can be removed, resulting in a cavity that has high fidelity to the size and surface details of the pattern.

2.3 The Pouring Channels

An important aspect after the cavity is generated is to design a system of vertical pouring channels (sprues and risers) and horizontal channels known as the runners. This process has to be done delicately to keep the casting sand intact while opening these channels with clay tools such as sharp chisels and knives and also round rods and pipes for the risers and sprues. A pouring basin can also be formed at the top of the main pouring sprue to facilitate the alignment of the molten metal with the pour. The horizontal channels, or risers, ensure that the molten metal would reach the cavity, then leave it into a riser. Once the molten metal shows from the riser, the pouring can stop and the metallic part can be recovered. Risers allow air to be pushed spontaneously from the mould cavity, and feed the cavity with metal during cooling down, thus mitigating the negative effects of shrinkage. For the same reason, a blind riser can be added to help with such metal feed. For larger moulds, vents can be added within the sand to allow air to leave and assist in cooling. All these details are shown in Fig. 4.

2.4 Part Post-processing

After the metal pouring process, the part is recovered by destroying the sand mould and placing the sand in a place where it can be recycled. It is clear that all the channels and gating systems that allow the molten metal to reach the cavity are themselves now filled with the metal. Figure 5a shows the part after cleaning the sand off, with all parts marked in red in Fig. 5b having to be sawn off and removed to obtain the desired part shown in Fig. 1a.

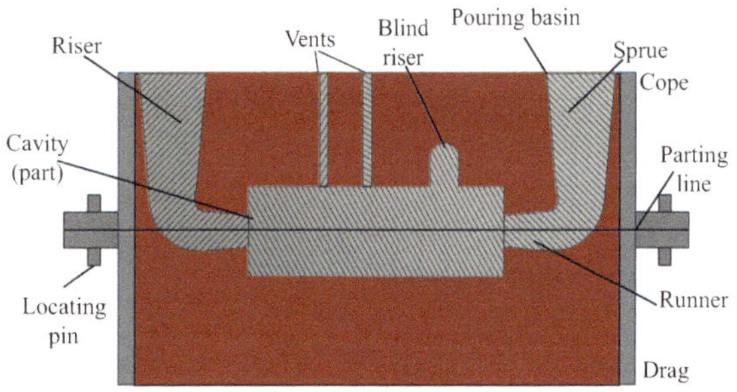

Fig. 4 Details cross-section of a sand casting mould with cavities and pouring channels

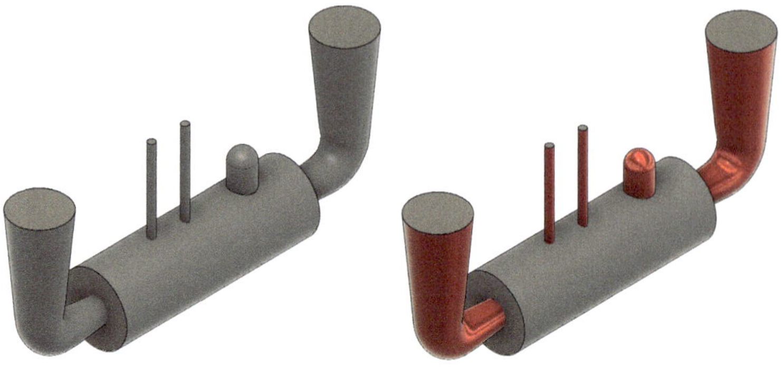
(a) (b)

Fig. 5 **a** the cast part recovered from casting sand and **b** parts colored in red must be cut off and removed

Table 1 Casting defects and proposed solutions

Defect	Definition	Cause	Remedy
Cold shut	Two molten metal streams solidify before fusing into a single shape	Multiple runners/sprues where molten metal is poured	Design the gating system to have a single inlet/unidirectional pouring
Shrinkage defects	Decrease in volumetric dimensions of the produced metal part	Metal intrinsic shrinkage during/after solidification	Add 2–5% extra dimensions as shrinkage compensation
Incomplete cast	The molten metal has not reached the whole cavity	– Temperature of metal melt is too low – Air trapped within cavity	– Increase pouring temperature – Add vents to the cavity
Flash and burs	Metal protuberances at edges of the metal part	Insufficient hold-down pressure on cope and drag allows the metal to overflow and solidify	Add weights/bolts to tightly secure cope and drag
Porosity	Produced part has a sponge-like structure	Moisture or air trapped within the mould	Dry mould surface with a torch or ensure adequate desiccation of the surface
Warping	The shape of the product is deformed	Premature part removal from the mould (while part is still hot)	Allow the part to solidify before removing from sand
Shift/mismatch	There is a translation/rotation of product at the parting line	Misalignment while gluing pattern parts during mould preparation	Add locating pins

2.5 Casting Defects

There are several defects that result from any casting process that have to be mitigated by procedural enhancements and provisions. A summary is provided in Table 1.

2.6 The Casting Sand

The sand used for casting is often called green sand and is a mixture of different types of powder/granular materials that have high cohesive bonding properties. The sand quality and adhesion properties have strong correlations with the integrity of the mould that is essential for the quality of the produced castings. This is especially true for the surface properties and fidelity to the original pattern design and surface attributes. There are many types of sand available for the casting process, some are found naturally and others are synthetic. Foundry sand was a crucial component of flourishing lead, tin, copper and bronze casting processes, as its availability in certain geographic locations made casting more amenable to regions in China, Persia, and Mesopotamia. Natural green sand is made up mostly of pure silica, around 9–10% bentonite clay that acts as a binder and the rest can be a carbonaceous mixture to promote a good surface finish of the castings. Water is also added to enhance the initial binding of the sand (Foundry Sand—Material Description—User Guidelines for Waste and Byproduct Materials in Pavement Construction—FHWA-RD-97-148, n.d.).

Synthetic sand is also commercially available and known as Petrobond, where the adhesive agent is made of oil. It is used to produce precision castings and will be presented in the DIY section of the chapter.

3 The Metal Melting and Pouring Process

The melting process of metals depends on both the sensible thermal energy needed to raise the temperature of the material and the latent thermal energy required to melt (energy of fusion) the material by breaking the molecular bonds and converting the material from solid to liquid. The liquefaction of the material allows it to flow and fill up the cavity left by the pattern within the die. The hotter the material is, the better the fidelity with which the molten metal will replicate the design features. On the other hand, pouring the metal at a higher temperature means that more energy has been used to keep the material above its melting point (since it is also its solidification temperature on the cooling side). This also means that the material will take longer to solidify which increases the processing time.

The melting process can take place in many ways and most of them require a huge energy and emissions footprint.

4 Electrical and Induction Furnaces

Using an electric or an induction furnace to melt nonferrous metals is a neat process in terms of the physical footprint and necessary provisions and precautions. The input power for such furnaces is electricity from the grid that is converted into sensible thermal energy via electrical current passage through electrical wires/coils. The main difference between electrical and induction is that the latter induces an electrical current in conductive materials that generates heat that would be enough to melt the material. Electrical furnaces are more basic and utilize the heat generated in the coils to melt the material (more electrical current usually results in higher temperatures in the furnace). The two varieties are shown in Fig. 6.

One of the main advantages of these furnaces is the ability to deploy them in well-ventilated indoor setups, and that they fit desktop spaces due to their compactness. Provisions have to be made to keep the areas around them clear, as their surface temperature can get high. However, the protective mesh screen around them mitigates the consequences of accidental contact with the furnace body.

The other, rather important, advantage of such furnaces is their programmability. A heating program that ramps the temperature up or down can be deployed for polymer burnouts (as will be discussed shortly) or to control the cooling to yield a specific metallurgical microstructure at quasi-equilibrium heating/cooling conditions. The burnout cycle can have a duration of 14–15 h, with ramp up, hold down, and ramp down over extended periods of time (hours) as shown in Fig. 7. Having a dedicated and accurate controller is essential for the success of this process, and the only furnace class that can handle such a task is the electrical one. Safety considerations are always recommended for such prolonged operations, such as proper exhaust provisions, using a webcam to monitor the process, and sending warnings in case of any unforeseen emergencies.

5 Gas Fired Furnaces

These furnaces are chemical ones where direct combustion of a combustible gas such as liquified petroleum gas (LPG), acetylene, butane, or others takes place. The oxidation usually is provided by supplying a pressurized air stream, or by utilizing the pressurized cylinders that contain the combustible gas directly. The generated flame (after a spark) is directed into the quadrant point of the wall of a circular furnace body creating a cyclonic effect as shown in Fig. 8. This effect raises the temperature of the furnace to the melting temperature of the metals (mostly nonferrous metals) that are usually placed in a crucible within the furnace. The crucible is made from a material that will not contaminate the metal being melted, such as graphite or quartz. Graphite crucibles are mostly used as although they are brittle, they are still tough and would sublimate (go directly from carbon to carbon dioxide) and will not unintentionally add contaminants to the molten metal. The process has many advantages, such as being a standalone process that requires no electrical provisions except for a supply of gas cylinders. The process is also economical and scalable and can be used on a modular scale at homes to seamlessly recycle aluminum cans, brass fittings, or copper pipes.

Although the process directly produces carbon dioxide, the gaseous nature of the fuel allows that to be the only harmful reaction product (water is the other one) as can be seen in Eq. (1) which shows the reaction between butane and oxygen:

$$C_4H_{10} + 9/2O_2 \rightarrow 4CO_2 + 5H_2O \tag{1}$$

From the equation, it can be deduced that only 58.12 g of butane is required to react with 144 g of oxygen in the air to provide the required heat to melt one kg of aluminum.

The amount of energy required to achieve the melting of the material in kJ/kg can be calculated from

$$E_{\min} = c_p.\Delta T + h_{fg} \tag{2}$$

where c_p is the specific heat of the material, ΔT is the temperature difference between the melting temperature and the ambient temperature while h_{fg} is the latent heat of fusion, or the amount of energy required for the phase change from

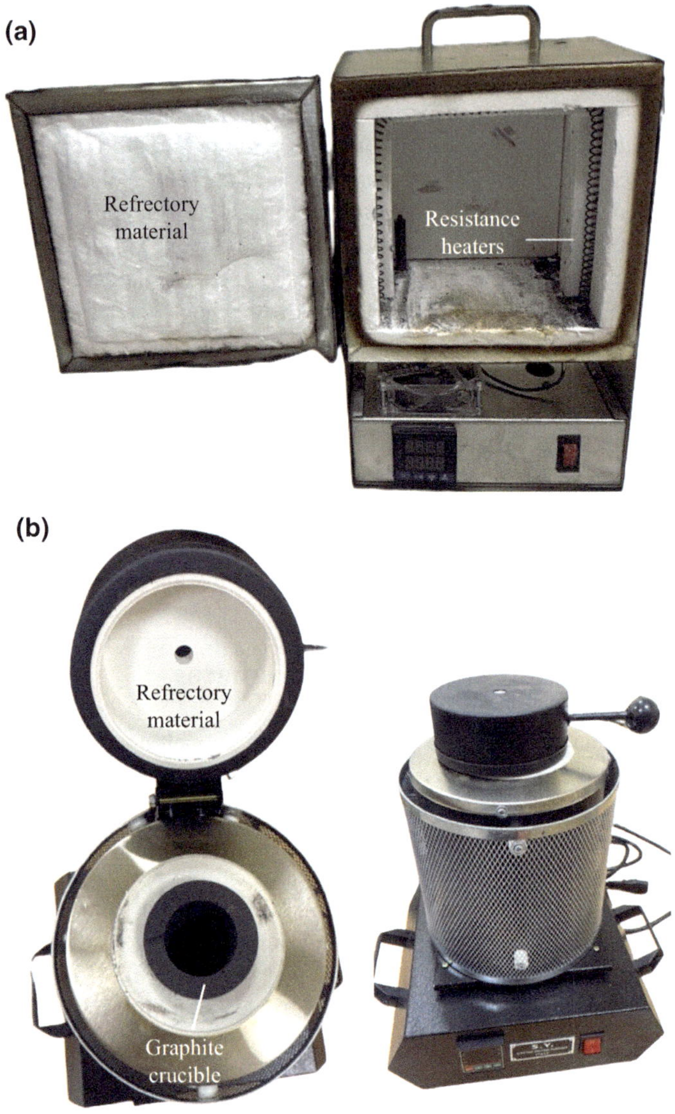

Fig. 6 **a** Electrical furnace with heating resistance and refractory interior, usually used for burnout processes, **b** and **c** compact electric furnace with graphite crucible for precious/nonferrous metal melting

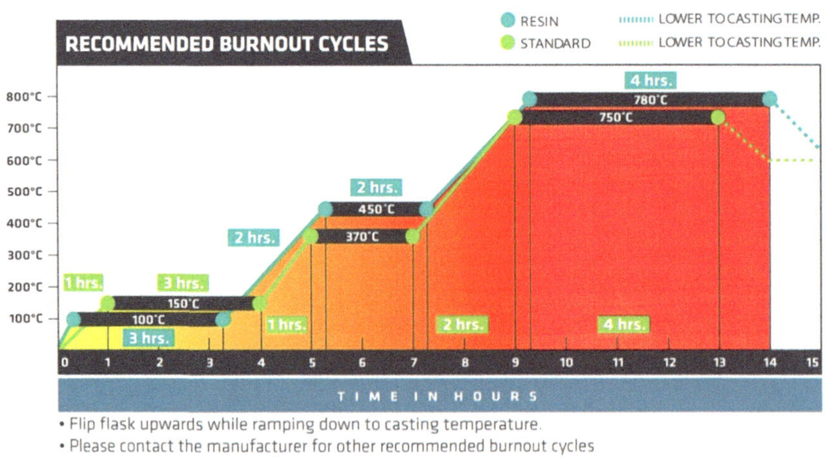

Fig. 7 Heat ramping process for burnout of PLA or resin materials (Certus PRESTIGE OPTIMA INVESTMENT POWDER, n.d.) (https://www.certus-int.com/optima)

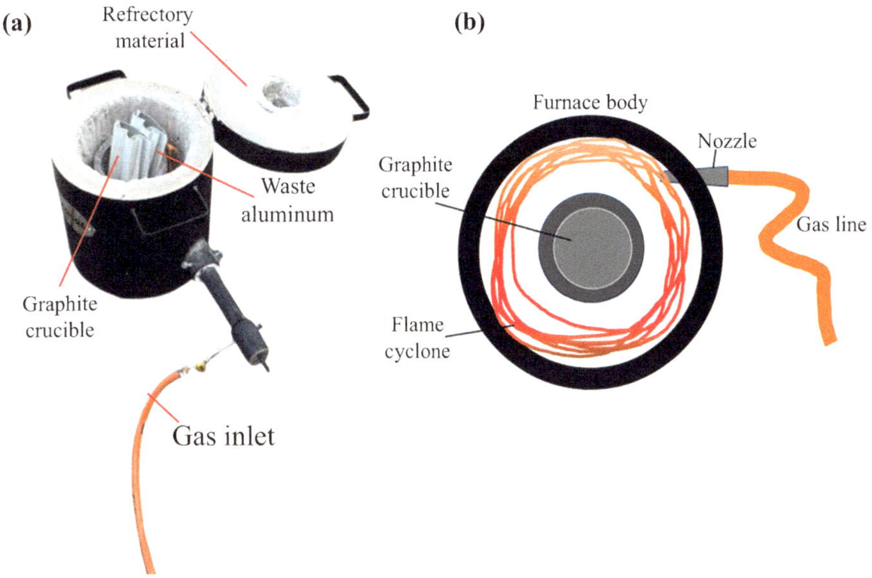

Fig. 8 **a** Gas-fired furnace and **b** the cyclonic flame effect within the furnace

solid to liquid. Different values of different materials are available in literature and textbooks on the thermal properties of materials. The resulting CO_2 gas can be captured and stored to mitigate the free escape to the atmosphere. The recycling process of a material such as aluminum is important, especially since nonferrous materials are easily recycled by this process given the astronomical increase in their prices over the past decade, more information can be obtained from (https://tradingeconomics.com/commodity/aluminum.). These trends show that once the price of these metals rises, it is virtually impossible for them to get back to their older price range even if the price per kilogram drop.

6 Solar Concentrators

Solar concentrators can be used to melt nonferrous materials, mainly using point concentrators. This would be the most environmentally benign process that would produce temperatures that are sufficient to convert solid materials into the required pourable liquid form. Temperatures at the focal point of the concentrators can reach up to thousands of degrees Celsius depending on the size (diameter) of the concentrator, as shown in Fig. 9.

This depends on the concentration factor, C, of an optical concentrator that can be calculated by relating the size of the solar aperture area (A_{ap}) of the paraboloid and the absorber area (A_{abs}) in the following formula:

$$C = A_{ap}/A_{abs} \tag{3}$$

The area of the absorber is the small area at the focal point where the light is collected from the paraboloid. The higher the concentration factor the higher the expected

Fig. 9 High temperatures at the focal point of the concentrating dish (Alami, 2014) (with permission no: 1511172–1)

temperature at the focal point. Although the process is independent, there are various challenges that potentially preclude applying it reliably. One of the issues is the convective and radiative cooling effects that increase with increased absorber temperature. This can be resolved by enclosing the absorber with a transparent container at the risk of losing much of the incident radiation as reflection losses as shown in Fig. 10.

There is also the time the process is available being limited to daytime and the more crucial factor of the

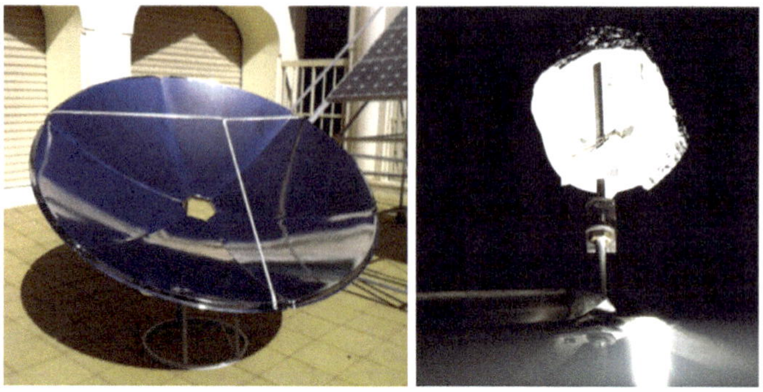

(a) (b)

Fig. 10 **a** Concentrating dish with secondary concentrator cover and **b** aluminum at the focal point (Alami, 2013) (with permission no: 5840131054879)

variability of the quality of the solar radiation. The process is extremely sensitive to the latter, as the clarity of the skies is a determining factor of the intensity of the incident radiation that otherwise would be scattered by suspended moisture, dust, aerosols, and/or other contaminants that adversely affect the quality of the required solar radiation to reach temperatures sufficient to melt metals. Point concentrators also require two-axis tracking, otherwise the focal point (point of maximum energy reception and thus temperature) will be away from the metal target. The melting process must have provisions to collect the molten metal which might become heavier with time and can interfere with the solar tracking process. There are point concentrators called beam-down concentrators, shown in Fig. 11, that have primary and secondary concentrators that allow for keeping the absorber static while the primary concentrators (or reflectors in the case of using a heliostat-type beam-down concentrator), which would allow an independent automated process to feed the waste metal (aluminum cans) and allow the molten metal to be collected and later solidify.

Fig. 11 Beam down concentrator at Masdar, Abu Dhabi, United Arab Emirates

7 Case Studies

The following paragraphs detail the conversion of design ideas into 3D-printed products that will consequently be made into metallic parts. Although 3D-printed polymeric parts can still be functional and be used as they are or assembled together to serve higher engineering functions, casting these parts out of metal will enhance their mechanical and thermal endurance, increase their lifetime, recyclability as well as allow a path other than material removal to be used to produce metallic parts. Although ferrous materials and alloys can be used for this purpose, their high melting points and complex smelting procedures preclude economic and environmental consciousness while recycling them for casting purposes. This is mainly the reason that nonferrous alloys that can be obtained from locally recycled everyday products (soft drink cans, copper coils, brass fittings, etc.) will be studied in this book. Two processes will be discussed, the first is the sand casting process and the other is the lost PLA process.

7.1 Sand Casting

The sand-casting process is used to produce an aluminum replica of a 3D-printed Moai (downloaded from thingiverse.com by Darthus, found at https://www.thingiverse.com/thing:149271) as shown in Fig. 12.

The model was sliced on Flashprint© software and printed without a raft on a FlashForge Adventurer 3. The model was printed as two halves in order to fit in the cope-drag setup of the casting system. The locating pins

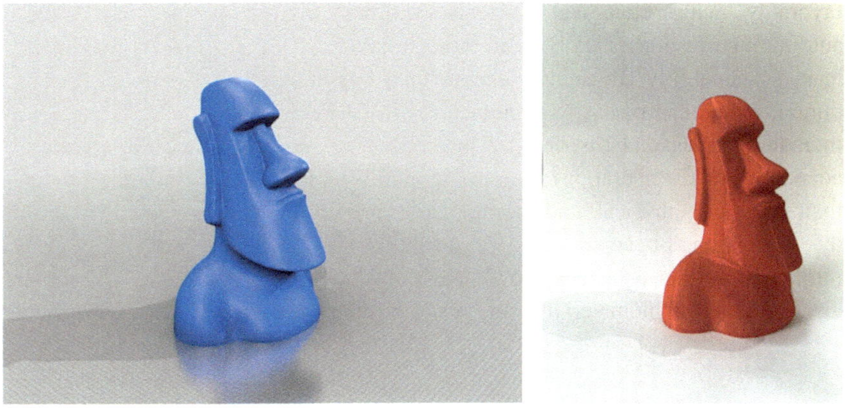

Fig. 12 **a** Moai model adapted from Darthus, licensed under CC BY-NC-SA and found at https://www.thingiverse.com/thing:149271 and **b** actual printed model

were not incorporated in the design as some of the casting defects could be highlighted later on Fig. 13 shows the printed parts from red 1.75 mm PLA.

The mould is made from steel (the cope and drag), although it could have also been made from wood or aluminum. The most important attributes to focus on are (a) flush contact between the cope and drag, (b) incorporation of locating pins to align the two halves, and (c) the model (pattern) orientation within the cope-drag setup.

The casting sand used is a synthetic Teton-Bond Petrobond sand, with a fine 130 Mesh Foundry Metal Casting Sand for High Detail, bought from desertcart.com. The drag is placed on a flat surface, as shown in Fig. 14a with the addition of a lubricant such as graphite or talcum powder to reduce friction between casting sand and pattern, which facilitates the removal of the pattern without damaging the resulting cavity. The sand around the pattern should be as fine as possible, so a sieve can be used to allow a high fidelity to contact as shown in Fig. 14. Later layers can be coarse but the sand has to be closely packed with a hammer, and the rest of the sand can be removed using a steel ruler to make the back surface as flat as possible. The drag is then flipped over and the pattern is now showing at the top of the drag as shown in Fig. 14c.

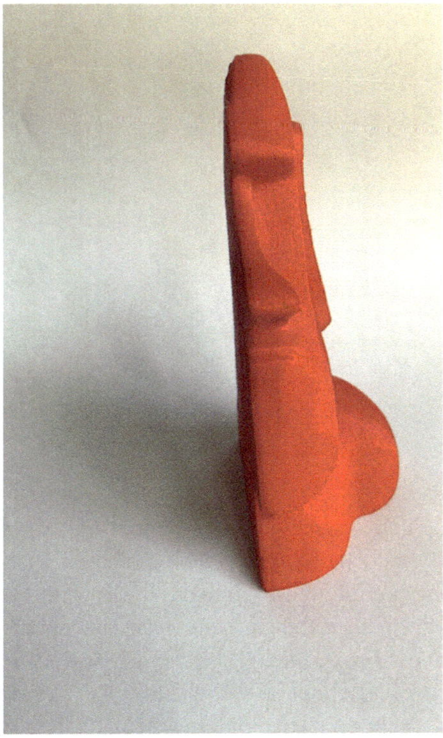

Fig. 13 3D-printed two halves of the Moai model

The next step is to meticulously attach/glue to other half of the pattern to the half enclosed in the sand, place the cope on top of the drag, and start filling and packing the sand, as shown in Fig. 15. At this stage, two 3D printed sprue and riser are added in locations where the pouring is expected to take place. Talcum powder is also added to facilitate pattern removal.

After the cope is filled with properly packed Petrobond sand, the runners can now be retracted, ensuring that this is done gently to preclude sand debris from falling into the cavity. The cope can now be slowly and carefully removed, exposing a cavity on one side and the full pattern on the other, as shown in Fig. 16.

It is advised that while working on preparing the mould that the furnace be fired and the material would start to melt. The mould preparation takes around 30 min (for a 30 × 20 cm mould) and about 2.4 kg of sand. The furnace firing usually starts with lighting a pilot flame using a torch and then opening the pressurized LPG or Butane container so that the pressurized fuel and air mixture would form a cyclonic flame as shown in Fig. 17b. The minimum required pressure to operate the furnace is 3 bar or above. Any lower pressures may impede flue gases from leaving the furnace chamber causing rather dangerous backfires. The flame should be orange in color and clear flue gases emanating

Fig. 14 Mould preparation steps **a** fine sand is usually placed around the pattern while **b** coarser sand then fills the rest of the mould half and is then compacted to arrive at the **c** final mould half ready for the next mould preparation step

Fig. 15 **a** Gluing the other half of the pattern, then **b** placing the cope through the locating pins on the drag and **c** adding the vertical gating system (pouring sprue and riser) to the mould

Fig. 16 **a** Two mould halves after sand-packing, and ready to open horizontal runner gates from the sprue impressions on the drag and then **b** removal of the pattern in preparation for metal pouring

from the circular opening, as shown in Fig. 17a. The scrap material is placed in an 8 kg graphite crucible since graphite is known to sublime at high temperatures and would not contaminate or stick to the melt. Occasionally, the surface of the metal should be sifted to remove contaminants that usually float on the surface, using a scraper or a stainless-steel spatula. The furnace can be temporarily switched off by switching off the gas supply to facilitate the removal. Then the melting is resumed until the metal has completely melted. The gas is then turned off again, and the crucible is carefully removed using tongs while wearing the safety gear. This includes a fire retardant apron and long gloves, safety boots, plexiglass visors as well as respirators.

The crucible is then tilted to allow the molten metal to pass through the sprue and into the mould, where the runners would allow the flow of the metal into the cavity and riser on

Fig. 17 **a** The operational furnace with flames emanating from the top orifice and **b** the removal of surface contaminants

the opposite side. Once the metal appears from the riser, the pouring operation should stop, and the crucible should be replaced within the furnace to avoid detrimental heat shocks that could result from leaving the crucible in the open and subjected to wind cooling or moisture. It is advisable to have thermal bricks to prop the mould on and a stainless-steel tray to contain any accidental spillages. Figure 18a shows the freshly poured metal in the mould, while Fig. 18b shows the opened mould with the cast part cooling down. It is noticed that since Petrobond has high oil content, there might be localized fires, and the inside of the mould cavity being scorched black. This is quite normal, and usually, the scorched parts can be removed, and the rest of the clay can be reused again. It is worth mentioning that a clear sign that the pouring process was a success is the molten metal showing from the riser, which indicates the free flow within the mould and the cavities being properly filled with metal.

After allowing 15–20 min for the part to cool enough to handle, the part is extracted and inspected. The part is freshly removed from the sand, with the runner system still attached, which will need to be sawed off later, as seen in Fig. 19.

The finished part is then inspected, and it can be seen that some defects exist. There is a certain amount of flash, which is an indication of insufficient holding down pressure on the mould halves, there sporadic pitting due to inclusions in the sand, and some extra metal accumulating in two corners where the pressure from the flowing metal has caused some of the sand to be displaced. Most of these defects can be fixed using a Dremel with a grinding bit.

One significant defect that was intentionally included is the slight misalignment. Since the two halves of the pattern did not have locating pins, gluing was not enough to keep them aligned and flush. There was some rotational motion during the packing of the sand in the cope that was transferred to the finished shape, as can be seen in Fig. 20. Thus, it is quite important to incorporate additional items in the design of the pattern, such as locating pins, to ensure that there are no unforeseen, and quite avoidable, defects transferring into the final part.

Fig. 18 **a** The mould cooling down after the pouring process and **b** the mould halves are open and the finished part is showing

Fig. 19 Metallic part removed from the sand and ready for post-processing, which includes removal of metal gates, surface cleaning

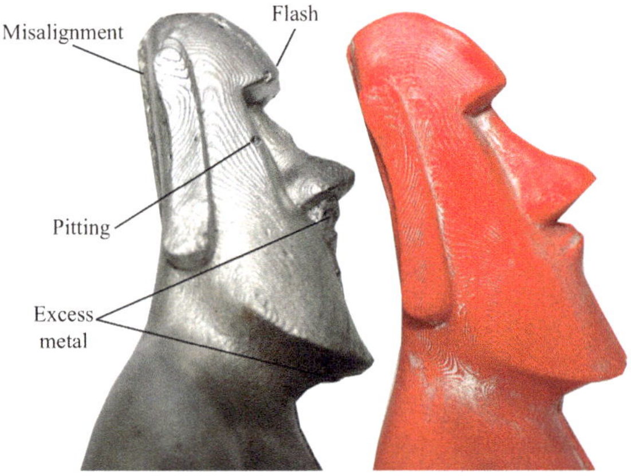

Fig. 20 Finished part (left) compared with pattern (right): it can be seen that there are several defects, but the surface fidelity to the pattern is exceptional, with 3D printing patterns showing on the finished part

7.2 Lost PLA Casting

The second method to design, 3D print and eventually produce metallic artifacts from PLA (or resin) material is the lost PLA casting. It is akin to the lost wax method that is primarily used to manufacture jewelry, and the equipment introduced in this section will demonstrate the utilization of 3D modeling and 3D printing in producing rings that have intricate surface details. The reader can experiment with different products, but this process is most suited for complex designs that are to be made from precious metals, as the equipment can be deployed indoors, and hence the manufacturing process has an intrinsic hermetic attribute. The process starts with the CAD model, then the 3D printed model, or models. Given the restrictions of the burnout plaster mould, the products can be stacked as a tree in order to maximize the throughput of the process, since it is a batch-type process. The following sections will take the reader through an experimental design and manufacture of a ring made from recycled aluminum. A single product such as a ring will be a part of a product tree, the design of which entails operational aspects that will be detailed as well.

(a) Ring Modeling and 3D Printing

The modeling process of the ring was done on Fusion360© and the design and modeling details can be found on this link (https://www.youtube.com/watch?v=zIkO0V6oAa4). The process took a sphere through several cutting processes to achieve the final design shown in Fig. 21 as side engravings were done. The top part of the ring could either have a stone fitted or another engraved matter. The design also boasts a cylindrical holder that will be fitted on the product tree later.

The ring file is then exported as a .stl file to be processed within the 3D printer-specific printing software. In the current case, FlashPrint© from FlashForge© is used, where supports are added, along with other parameters such as filament density, build temperature, nozzle speed, and others. Since only the shell is important in the lost PLA application, the infill was chosen to be 20% only, as this saves printing time and also requires less energy and time during the burnout phase. These steps are shown in Fig. 22.

Since the mould making part of the lost PLA process is in general material intensive, it is always advisable to prepare the products in bulk and on a product tree. More than one

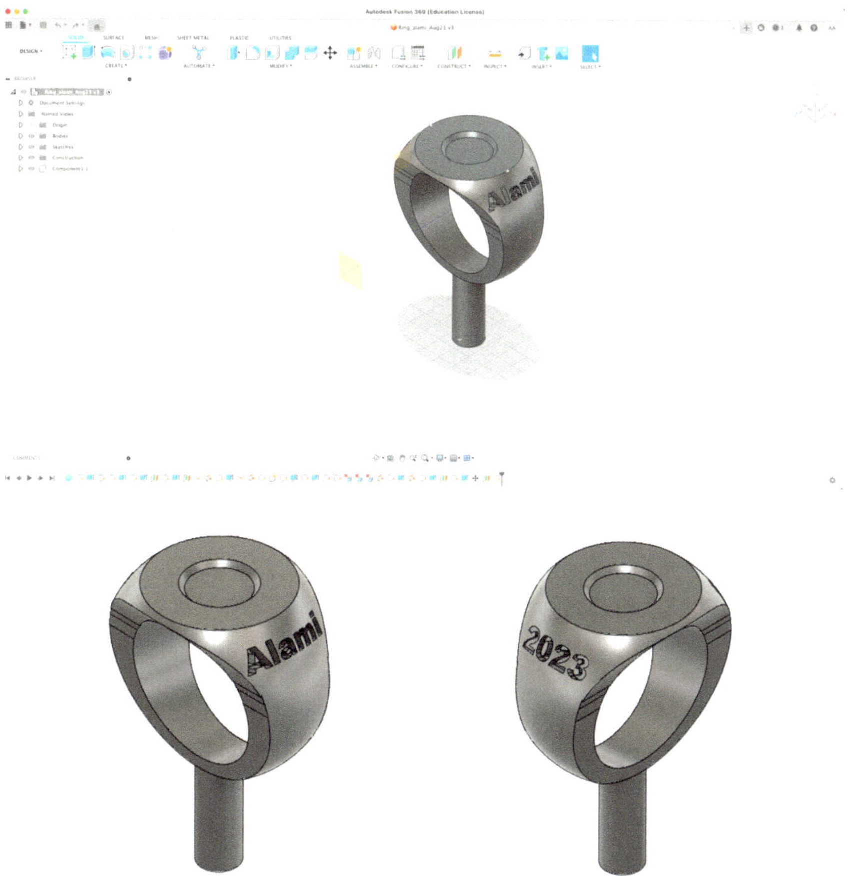

Fig. 21 3D model of the ring constructed in Fusion360©

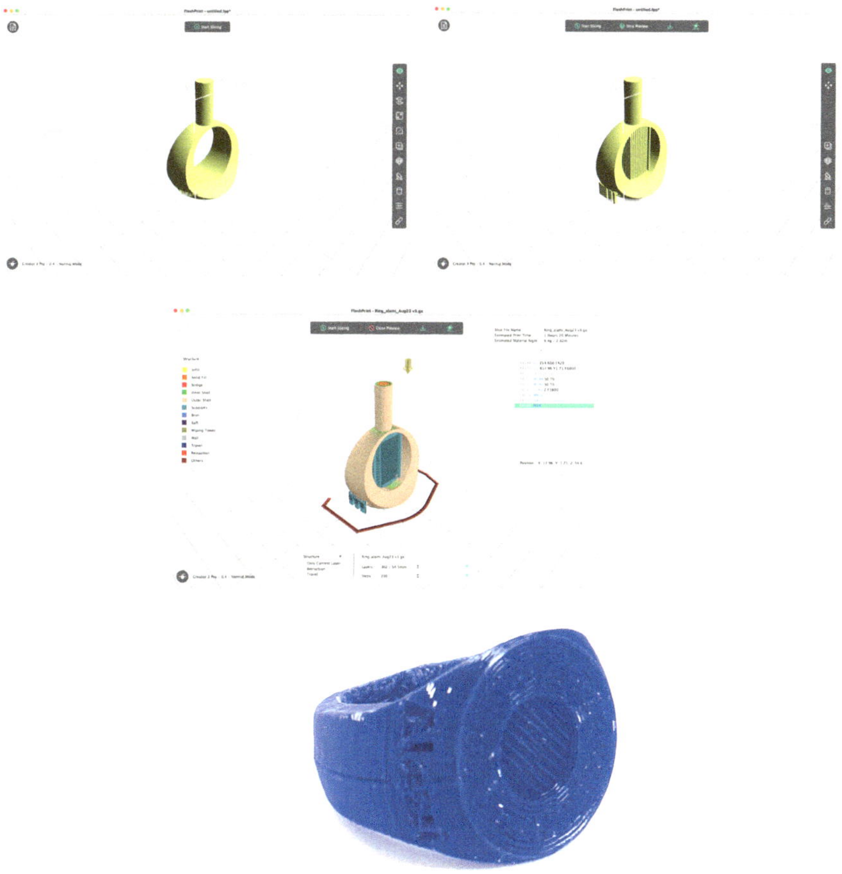

Fig. 22 .stl file preparation in FlashPrint© software to be sent to a 3D printer. The printed ring is shown in the lower right corner

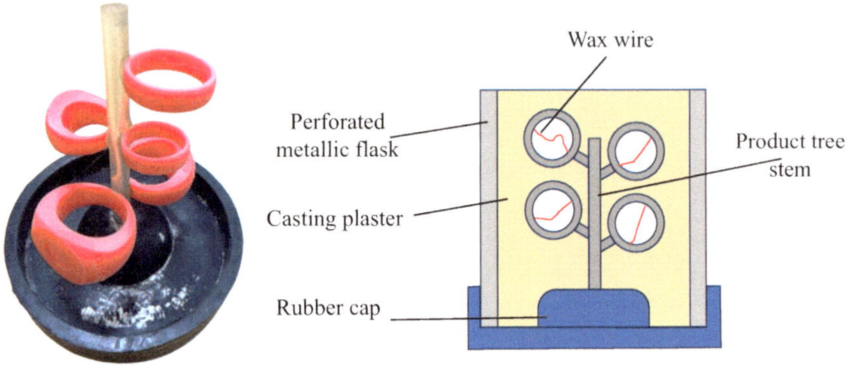

Fig. 23 Product tree of multiple rings bonded by wax on a stem supported by the silicon rubber cover of the mould. Wax wires are added to allow internal channels of metal flow and to release air from the cavity

ring was printed for this process and all of them were placed on a stem as shown in Fig. 23. Wax wires can also be added to allow internal channels of metal flow and to release air from the cavity.

(b) Mould Preparation

A special plaster material (Prestige ORO 5Lb Premium Jewelry Making Casting Powder) is used to host the mould cavity. The preparation procedure calls for thoroughly mixing 1000 g of the plaster with 480 ml of distilled water using a kitchen blender, then placing the mix in a vacuum bell to deaerate it for 5 min as shown in Fig. 24. The investment casting apparatus has two sections, one is a sealed vacuum dome that helps remove dissolved air bubbles from the plaster and the other section is a vacuum casting chamber, where the vacuum pump in the apparatus is used to provide a negative pressure on the dry, porous plaster that sucks the metal into all parts of the cavity, allowing high fidelity surface reproduction.

Once the plaster is deaerated, it is quickly poured into the perforated, but temporarily taped up, casting flask that has the product tree placed within, and plugged from one side by the silicon rubber seal, as shown in Fig. 25. The tape allows the plaster to be poured and allowed to dry without leaking from the flask. The perforation in the flask wall allows the utilization of the negative pressure in the casting chamber during the casting phase.

The flask is then placed once more in the vacuum chamber to remove air bubbles for 20–30 min while the plaster slowly dries. The flask is then removed from the vacuum chamber and the silicon rubber seal is also removed, revealing the stem of the product tree, as shown in Fig. 26a and b, respectively.

The flask now is ready for the burnout cycle that will last for up to 15 h as was shown in Fig. 7 in the oven shown in Fig. 6a. The orientation of the flask should be as to allow all PLA material to flow through the stem of the product tree, leaving a cavity of the exact same shape, as shown in Fig. 27. Proper safety precautions must be in place for this

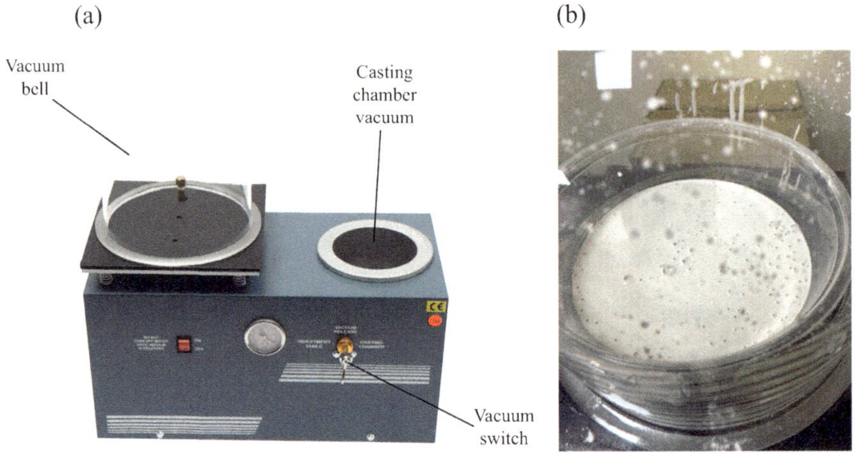

Fig. 24 **a** Investment casting apparatus with vacuum bell for deaerating casting plaster and **b** an image of air bubbles leaving the plaster

(a) (b) (c)

Fig. 25 **a** Product tree placed in casting flask, **b** plaster poured into the flask and placed again in the vacuum chamber to remove dissolved air bubbles and **c** the dried-up plaster

(a) (b)

Fig. 26 **a** Casting flask with the silicon rubber seal removed and **b** the stem of the product tree showing, ready for burnout cycle

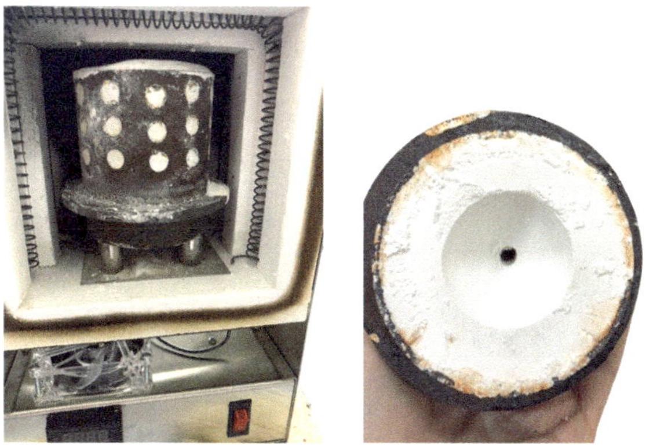

(a) (b)

Fig. 27 **a** The casting flask placed within the burnout furnace and **b** the flask after the burnout process

process, such as proper ventilation, furnace monitoring, etc. should be active in order to prevent any accidents. The furnace itself will get hot to touch, but its effective heat envelope was observed to be low, it is advisable to place the furnace away from any sensitive or combustible materials or objects.

(c) Casting

The next step is to place the flask within the vacuum casting chamber, shown in Fig. 28, and also melt the selected material (scrap aluminum in this example) using the induction furnace that was shown in Fig. 6b. The graphite crucible is capable of holding 2 kg of material, but if more material is needed, the gas-fired furnace can also be used, although obviously not in a lab. Proper gaskets are used underneath

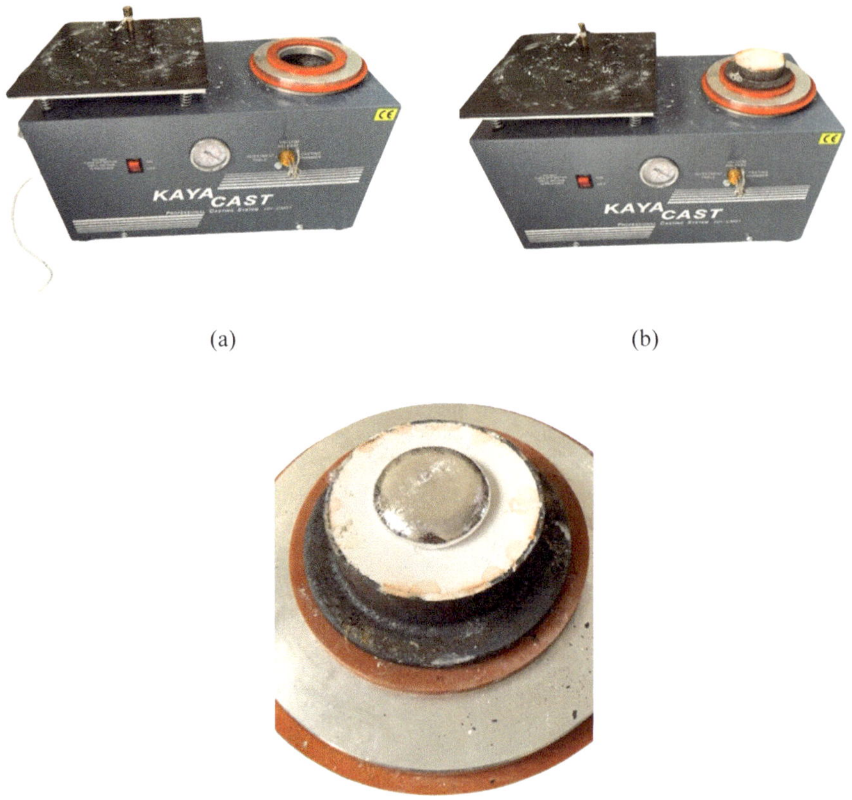

(a) (b)

(c)

Fig. 28 **a** The casting system with proper seals and gaskets, **b** casting flask placed in vacuum system and **c** the cast aluminum within the cavity

the flask to allow proper sealing for the vacuum to build and the metal to fill the cavity with the suction force of the negative pressure.

After the aluminum is allowed to solidify (2–5 min), the flask is placed in water and pressure washed to remove the plaster and reveal the cast product tree, shown in Fig. 29.

The final step is to clean the surface up with a wire brush and cut off each product and polish it with successive sandpaper grades as shown in Fig. 30.

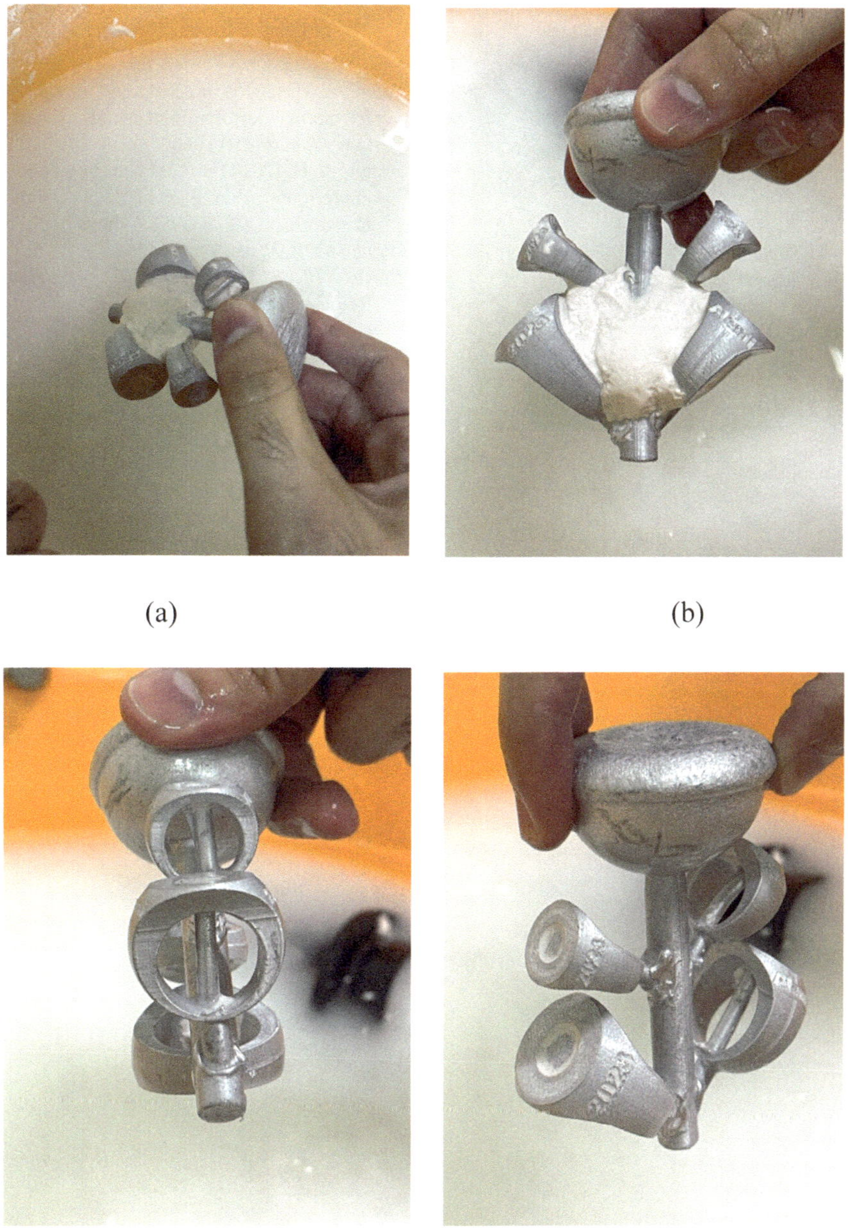

(a) (b) (c) (d)

Fig. 29 Cleaning steps of the product tree by dissolving it in water. It is interesting to see the surface details of the 3D printing process on the finished products

Fig. 30 Final product (unpolished)

References

Alami, A. H. (2013). Investigation in utilizing paraboloid concentrators for thermal perforation of metals. *Renewable Energy, 56*, 105–109. https://doi.org/10.1016/j.renene.2012.10.037

Alami, A. H. (2014). Assessment of using secondary concentrators for nonferrous material removal applications. *Advanced Materials Research, 939*, 506–513. https://doi.org/10.4028/www.scientific.net/AMR.939.506

Certus PRESTIGE OPTIMA INVESTMENT POWDER. (n.d.). Retrieved July 24, 2024, from https://www.certus-int.com/optima

Foundry Sand—Material Description—User Guidelines for Waste and Byproduct Materials in Pavement Construction—FHWA-RD-97-148. (n.d.). Retrieved August 1, 2024, from https://www.fhwa.dot.gov/publications/research/infrastructure/structures/97148/fs1.cfm

Santhi, S. (2018). Calculation of shrinkage of sand cast aluminium alloys. *International Journal of Applied Engineering Research, 13* (11), 8889–8893.

GPSR Compliance

The European Union's (EU) General Product Safety Regulation (GPSR) is a set of rules that requires consumer products to be safe and our obligations to ensure this.

If you have any concerns about our products, you can contact us on ProductSafety@springernature.com

In case Publisher is established outside the EU, the EU authorized representative is:

Springer Nature Customer Service Center GmbH
Europaplatz 3
69115 Heidelberg, Germany

Batch number: 10762593

Printed by Printforce, the Netherlands